Fishing, Foraging and Farming in the Bolivian Amazon

Lisa Ringhofer

Fishing, Foraging and Farming in the Bolivian Amazon

On a Local Society in Transition

Springer

Lisa Ringhofer
IFF, Institute of Social Ecology
Schottenfeldgasse 29
1070 Vienna
Austria
lisa_ringhofer@yahoo.co.uk

ISBN 978-90-481-3486-1 e-ISBN 978-90-481-3487-8
DOI 10.1007/978-90-481-3487-8
Springer Dordrecht Heidelberg London New York

Library of Congress Control Number: 2009942031

Printed on acid-free paper

Springer is part of Springer Science+Business Media (www.springer.com)

Preface

This book presents the résumé of intensive research of the past 5 years. It is an extension of my PhD thesis, but now placed in the wider context of transition studies. Thematically, this monograph contributes the fourth local metabolic case within the field of local transition studies and may hence be regarded as a follow-up of the three previous local studies carried out. However, the book attempts to explore new territories, too, I hope, by integrating the use of human time into the existing biophysical framework.

The foundation for this book was already laid in 2001. I had just come back from a rural development work assignment in Guatemala, an experience which evoked my interest for more theoretical grounding on sustainability of society–nature interactions. The Institute of Latin American Studies at London University proved just the right place to be at the time. Inspired by the intellectual milieu, it was Professor Christian Brannstrom, whose indefatigable lectures on indigenous knowledge and strategies on sustainable resource management in the Amazon, I found most compelling. What I did feel, however, was a lack of specific methods and tools providing concrete indicators on the sustainability of specific resource management strategies; with the underlying idea of integrating these instruments into my world of development projects. I knew then that to this end, I would have to leave the secluded world of social sciences and start interacting more consistently with natural science approaches.

But this was not until my return to Vienna in early 2003, when I literally stumbled into a lecture on the social metabolism of societies held by Marina Fischer-Kowalski, Professor at the Institute of Social Ecology in Vienna. What sounded rather peculiar at first, did, however, provoke my interest for more. To me, the borrowing of an originally biological concept made sense insofar as it shows the crucial dependencies between the different biophysical resources that are exchanged between a society and its natural environment. This I found to be particularly interesting when applied to traditional societies, where dominant modes of production are based on biotic technologies (e.g., hunting and gathering). This initial intellectual fecundation and the subsequent immersion into the existing publications on this topic led to the birth of the core of this book: an analysis of the metabolic relations of a local community in the Bolivian Amazon, a study that was encouraged wholeheartedly by Marina Fischer-Kowalski and some other forthcoming researchers at the Institute. For a young student setting out to explore the realities of fieldwork for the first time, the idea of spending some time with indigenous people in the Amazon – a place that evoked an image of unplumbed mysterious depth – was indeed thrilling.

As I was preparing myself for my forthcoming fieldwork sojourn, one day during an informal discussion in her office, Marina Fischer-Kowalski shared with me her thoughts on how to integrate human time use into the existing metabolic framework. She sketched her ideas on paper and gave me some useful indications on how to proceed from there. In hindsight, my attempts to bridge these ideas with more common classifications of time use and apply them in an empirical field situation, were probably the most challenging, but also stimulating, parts of my entire research project. The results of this endeavour are presented in Chapter 5 of this monograph.

Admittedly, to some scholars this book may be a too much personally shaped narrative. Although what I report in this book is an effort at detached description, based on an explicit methodological framework, in appropriate places in this monograph I have included my own thoughts and provided examples of real persons in order to exemplify my point. This rather personal approach, so I believe, also eased my guilt for reducing the 'real' life of a community into a set of quantitative indicators. Where suitable, I have also included lengthy qualitative discussions on the symbolic aspects that shape the interactions of the Tsimane' with their natural environment. I hope that some readers have appreciation for this narrative style, finding the account enjoyable as well as informative, and in the best of worlds, both of these at once.

Personal names used in this book are in most cases actual names and sometimes pseudonyms. Actual names are used for persons who have given their permission and for persons whose portrayal in the text is non-problematic. This reflects the fact that the Tsimane' of Campo Bello are generally pleased to have their real identities represented to a larger world. Pseudonyms have been used in the rare cases in which the information could be perceived or interpreted as unflattering. A final note on naming relates to the use of language. At various text passages throughout the monograph, the reader finds words either written in Tsimane' or Spanish. All these terms are written in italics (e.g., *chätidye'* or *chicha*). Likewise, when a Tsimane' word emerges for the first time, the reader will always find its corresponding English translation in parentheses, e.g., *chätidye'* (relative or kin). The scientific term for flora and fauna species will also be written italicised (e.g., *Swietenia macrophylla*).

Vienna, May 2009 Lisa Ringhofer

Acknowledgements

Any enterprise of this scope depends on the accumulation of knowledge of many people in different parts of the world. While I am not able to acknowledge all those individuals who have, knowingly or unknowingly, influenced or contributed otherwise to the making of this book, I would still like to dedicate a special 'thank you' to those individuals upon whom so much of this book has depended.

First of all, since the people of Campo Bello – their homes, gardens and the forest – constitute both the starting point and the end point of my inquiry, the present book is not least dedicated to all of them. That I had the opportunity to attain my research goals and find inspiration in their surroundings owed most to their generosity and tolerance; they were good mentors and patient hosts throughout my stay in their community. Out of sight, but close to heart and mind, my gratitude goes particularly to the local teacher Zacarías Miro, simply known by everyone as the *profe*. He and his wife Juanita provided much valuable support and dedication throughout my stay and introduced me slowly to their world. I also have in mind Roman Durvano and Iguiño Pache, both my patient translators and reliable guides throughout the seemingly endless rounds of data compilation. To the children of Campo Bello, especially Mario and Elena, I am indebted for having guided my first steps in the forest, where I was brought closer the real meaning of *sobaqui*. I would like to thank you for that beautiful gesture. Needless to say, there have been several other Tsimane' informants, whom I made encounters with at various time and locations; each has enriched my knowledge base and broadened my horizon.

In San Borja, I think with particular fondness of Doña Dalia and her family who provided a nice home where I could come back to every so often. I am also grateful to the members of the *Gran Consejo Tsimane'* for having granted me a permit to conduct research in their territory. Without the support of this native organisation, my project would have never become reality. I also owe a great deal of cultural guidance to Tomás Huanca and Ester Conde, who both share my interest in the Tsimane' culture and were the best of companions most of the evenings in San Borja. I treasure memories of my conversations with both of them, which gave way to a still lasting friendship. The municipality of San Borja and the directorate of the Beni biosphere reserve based in San Borja also helped me over the course of my fieldwork. The departmental research centre CIDDEBENI in

Trinidad granted me access to an extensive library and provided me with important data sources that were required for the purpose of the biophysical analysis. My gratitude goes to all of them.

My entire research endeavour was inspired and guided by the Institute of Social Ecology in Vienna, with my study being embedded in the Institute's large-scale 'Local Transition Studies' programme. I was welcomed to use the resources and infrastructure of the Institute to which I am still provided generous access to. I would particularly like to emphasise the support of my 'doctoral mother' Marina Fischer-Kowalski who retained a stalwart belief in my academic credibility throughout the entire course of research. I am also indebted to her for the possibility to make my work accessible to a larger scientific community beyond the confines of the university setting. My colleague and *amigo* Simron Jit Singh deserves a special mention, as he has been my mentor in many aspects. In this context though, I am really grateful to him for having taught me that adding a personal touch to scientific writing by no means reduces the quality, but enriches the reading experience. Also, he has listened with unflagging good humour to my doubts and critical reflections on sustainability and development. Clemens Grünbühel I would like to thank for his more indirect, but just as significant support, as his writings on the communities of SangSaeng and Nalang have inspired my own work in many ways. Most importantly though, as he was the first researcher to dare the descent into the depths of local-level metabolic empiricism, the merit goes to him for being first in solving the variety of methodological challenges that were encountered in the field. As a veritable novice in the thickets of metabolic literature at first, his compellingly detailed descriptions surely prevented me from any large blunders. In contrast to the diversity of fieldwork, I saw data analysis and writing as a lonely enterprise, which I conducted in self-absorbing solitude. In this respect, I am grateful to Heinz Schandl who pulled me out of my thoughts every so often to offer his helping hand with tricky biophysical calculations. Helmut Haberl, Karlheinz Erb, Fridolin Krausmann and Willi Haas played a more subtle yet significant role in encouraging my research endeavour. None of these mentors are, of course, responsible for my errors or deformations that may have occurred in the thought of those who helped my research endeavour.

I would also like to express my warm thanks to Maria-Anna Edlbacher and Martin Hochmeister for helping me with the graphs and illustrations. To Scott Hayes, I am grateful for his careful proof-reading and valuable comments on improving the language of this book. Thanks to all three of you for your timely support. For the often invisible work and skill that transform an author's rough manuscript into a published book, I am grateful to Springer editors, who were supportive of this project from the very beginning and who ensured that the book made its way into print in a timely manner. Particularly I should like to thank Takeesha Moerland-Torpey for her guidance on preparing the manuscript; her valuable comments were appreciated throughout.

Finally, my gratitude goes to my family. My father's mathematical skills and his patience have saved me hours in despair. Thanks to my mother's provision of culinary delights during the time of writing, I was able to wholly concentrate on my study. I wish to also thank my sister and friends for providing the mostly moral support I somewhat needed for finishing the manuscript. None of this, however, would have been possible without the support of my partner throughout this voyage, Andreas, who has never lost faith in me and was relentless in his intellectual and emotional encouragement.

Contents

List of Figures

List of Tables

Chapter 1
Introduction: Setting the Stage

Abstract This chapter outlines the overall aims of this monograph and introduces the reader to the conceptual and empirical setting. Here, I will present a short introduction to sociometabolic regime transitions and the MEFA framework, both of which provide the conceptual base for the subsequent chapters. Moreover, this introductory chapter also includes a personal account of the empirical fieldwork situation, discussing the journey from first encounter with biophysical studies to life in the village. Finally, I will discuss the main outline of the book, describing the contents of each of the following chapters individually.

1.1 Introduction

We have just lived through a fast-changing century. During the twentieth century, the world population quadrupled, the average human lifespan doubled and the global economy increased at enormous speed and intensity levels, accompanied by rising levels of material and energy consumption and carbon dioxide emissions. The outcome of these growth levels has been a deterioration of many global environmental performance indicators in recent decades and the risk of transmitting a negative environmental legacy to future generations. So, at the beginning of the twenty-first century, we stand at a crossroads not previously encountered as it seems that the path of progress and change is no longer unending. These concerns, shared by many, have shaped a whole research community to illuminate, from different angles and within different time frames, the notion of a Sustainability Transition. If we consider a transition to be a condition that lies between two qualitatively different states, implying a rather dynamic and probably chaotic process of change with yet an unclear outcome (Fischer-Kowalski and Haberl 2007: 3), then this also implies that a Global Sustainability Transition differs from previous transitions (e.g., the demographic transition) insofar as it requires deliberate human intervention to make it happen (Moran 2006; Parris and Kates 2003). But the global is inherently connected to the local. Transitions do not happen simultaneously but change, however, tends to radiate gradually from the centres of novelty.

L. Ringhofer, *Fishing, Foraging and Farming in the Bolivian Amazon: On a Local Society in Transition,* DOI 10.1007/978-90-481-3487-8_1,

In this respect, more traditional societies still survive in places that are physically remote and culturally isolated (Raskin et al. 2007). It is estimated that today, more than half of the world's rural population still lives remotely under traditional, agrarian conditions, mostly in the so-called developing countries. However, also these societies are currently undergoing processes of change, as local systems are becoming increasingly more enmeshed with wider influences and larger regions. Many developments have triggered cultural and social changes, leading to more active engagement with national and global lifestyles. We may not forget that these transition processes 'at the periphery' are much more fundamental than those of the industrialised 'centre', since their still relatively low levels of resource consumption imply an enormous future growth potential for material and energy consumption (Schandl and Grünbühel 2005).

For a Sustainability Transition to happen unanimously at all levels, from local to global, it is important that sustainability is pursued in ways that allow the needs of these local non-industrialised societies to make gains in their material conditions of living. However, the prevalent development model propagated by numerous development programmes somewhat contradicts this sustainability notion. Supporting the integration of remote communities into the market economy through the promotion of certain cash crops, for example, is likely to promote an industrial development approach that entails an increase in the use of fossil fuels, either directly (through the industrialisation of their agriculture) or through the use of more indirect agricultural inputs (e.g., specialised machinery). These current trends though run counter to the broad evidence we are provided with by global change research (see, for example, Schellnhuber 1999); namely that we are heading straight towards the erosion of our own natural resource base. A transition towards a more sustainable global future, on the other hand, would require a profound change in consumption patterns away from those currently prevalent in the industrial north to more modest levels (Fischer-Kowalski and Haberl 2007; Wackernagel and Rees 1996).

So, if current development patterns are to be guided away from existing 'fossil-fuel-based approaches', there is an inherent need to first of all gain an in-depth understanding of the transformative potential and impacts that are generated at the society–nature interface when processes of change get underway. This is what this book is concerned with. Through the lens of the local, it explores the biophysical structures and process dynamics of a local rural community under transition. Set in the Bolivian Amazon, the study examines the society's material and energy relations with its natural environment; all this in an attempt to identify the overall pressure on the environment (through material, energy and land use) as well as the burden on the people (through labour time). Its specific aims are twofold: first, a 'snapshot' on the biophysical relations of the indigenous Tsimane' with their environment taken at one moment of their course of development should help modelling sustainable development trajectories for larger regions. No less important and inherently connected to the first is my second interest. Based on the system's biophysical resource profile, what are the local opportunities, challenges and constraints the village of

Campo Bello is currently facing on its sustainability transition pathway? In general, I hope that this book will incite constructive debates about contemporary approaches to sustainability, as it provides quantitative indicators that measure development directions. It may be of interest for both development practitioners and researchers of ecological economics, environmental and natural resource sociology, cultural and ecological anthropology, human population–environment dynamics and international rural development.

1.2 Introducing Society, Nature and Transitions

The idea for the empirical study of this book has grown out of a concerted effort between a team of social and natural scientists at the Institute of Social Ecology in Vienna, whose theoretical interest lies in linking local transitions with the global arena. As the concept of sociometabolic regime transitions will be elaborated in detail in Chapter 2, for now, we may contend ourselves with its central propositions only. The sociometabolic transition concept looks at common development trajectories towards possible and impossible futures from a biophysical perspective. To this end, it offers a comprehensive methodological and conceptual toolkit for analysing society–nature interactions. Within this paradigm, approaches thrive to understand the structures and dynamics of a social system under transition (a national economy, a region or a local community) in order to determine future trends and plan suitable interventions in these social processes accordingly (Schandl and Grünbühel 2005).

The leitmotif that spans like a bow over the transition study paradigm, however, is that humans have the ability to transform nature, thereby triggering intended and unintended changes in their natural environment to which societies, in turn, react (Fischer-Kowalski and Haberl 2007). What is central to this assumption is the equal standing of society and nature. From a social science perspective, this is quite a novel approach as sociologists are generally concerned with the 'study of society' and hence inclined to term the environment as a human construct rather than a complex system of its own (Cudworth 2003: 2). However Maurice Godelier, probably one of the world's most distinguished Marxist anthropologists, plays a fundamental role in refining our understanding of the interconnectedness between societies and nature. In his book *The Mental and the Material* (Godelier 1986) his aim is to explore the relations between thought, the economy and society. He stresses the importance of the material relations humans entertain with their natural environment and the transformative impact of a society's use of nature: '(the natural environment is) a reality which humanity transforms to a greater or lesser extent by its various ways of acting upon nature and appropriating its resources. Nevertheless an ecosystem is invariably a totality which is only reproduced within certain limits and which imposes on humanity several series of specific material constraints' (Godelier 1986: 28). He makes a distinction between three largely different ways of appropriating nature, starting with that part of nature which remains outside the direct or indirect control of humans but still affects everyday lives (e.g., climate). Secondly, he refers to the part of nature that is indirectly transformed (e.g., soil erosion). His final distinction centres on the direct transformation of nature by human beings that cannot be reproduced without their energy and labour input.

Yet I shall not linger over this point but look at the historical dynamics Godelier attributes to the interactions between humans and their natural environment. To his mind, 'people have a history because they have the ability to transform nature' (Godelier 1986: 1-2), a claim that has guided the research team at the Institute of Social Ecology throughout the process of shaping and refining the sociometabolic regime transition concept (Fischer-Kowalski and Haberl 2007). The central claim here is that it is the interaction process between a society and nature that shapes the development path of this society. This is indeed in stark contrast to the nineteenth century evolutionism writings of Charles Darwin or Herbert Spencer which present history as a linear continuation of a biological evolution, advancing from the lower to the higher. Transitions from a lower to a higher state, in their view, occur endogenously, due to built-in process dynamics with little or no attention to deliberate human agency as a motor to steer the developmental direction.

Godelier (1986: 108) also contends that each socio-economic system determines a specific mode of exploitation of natural resources and implementation of human labour-power. Interestingly, his analysis of various cases reveals that societies display certain elements in common, irrespective of the constraints imposed by specific biogeographical conditions but rather depending on the society's capacity (or incapacity) to act upon it. Sociometabolic regimes are equally characterised by common patterns of society–nature interactions, which are the exchange of energetic and material flows from and to the environment, the use of particular technologies and the transformation of natural systems through labour. Methodologically, this is achieved by applying the notion of 'social metabolism' and 'colonization',[1] both of which are conceptual tools for describing the interaction of social system dynamics with natural system dynamics; and both are subsumed under the MEFA framework.

1.3 Introducing the MEFA Framework

Conceptually and methodologically, the MEFA (materials and energy flow accounting) framework lies at the heart of analysing the exchange relations of material and energy flows between a social system and its natural environment, the feedback processes and the biophysical constraints of the particular system. The MEFA approach sheds light onto the biophysical dimensions of production and consumption

[1]The spelling of the term 'colonization' and all its derivatives has been taken from original publications and is hence spelt with a 'z' rather than with an 's'.

processes, thereby establishing a link between socio-economic variables and biophysical processes. MEFA equally provides a set of sustainability indicators that measure societal impact on natural processes and the social organisation of socio-economic modes of production. Within the MEFA approach there are two specific tools for describing society–nature interactions; these are termed as 'social metabolism', on the one hand, and 'colonization of terrestrial ecosystems', on the other hand (see Fischer-Kowalski and Weisz 1999; 2005). Social metabolism is based on the premise that a society not merely reproduces itself culturally, via communication (as argued by Luhmann 1984), but also biophysically (population, livestock and artefacts) by means of a steady exchange of matter and energy flows with nature as well as other societies. Different sociometabolic regimes display different metabolic profiles, which is the quantity and quality of matter and energy used for living. 'Colonization of terrestrial ecosystems' is the second concept employed for describing society–nature interaction. Besides the exchange of matter and energy between social and natural systems, colonization refers to the deliberate intervention of a society into a natural system with the intention of transforming certain parameters to render it more useful for the human population who exploits it. By so doing, natural systems are brought into an ever more vulnerable state, or if we use a term taken from transition concepts, a state far from dynamic equilibrium, and a continuous input of labour, energy and materials is needed for the natural system to remain in that colonized state.

MEFA has been effectively used for national level accounts to describe basic processes within the human environment metabolism and to compare economies with each other (see, for example, Eurostat 2007). Local MEFA studies, on the other hand, provide a 'snapshot' profile at the base of national economies in transition, thereby shedding some light onto the scale interaction between the national and the local stetting (Singh 2006). It is indeed at the local base where the actual impacts of policy implementation at higher scale levels can be observed at first hand. Admittedly, the application of MEFA at the local level is still less elaborated and only three case studies were conducted prior to my effort. It is hoped that this book contributes to refining the local level methodological toolkit.

The MEFA framework entails three interrelated sets of relations (Fischer-Kowalski and Haberl 2007: 17) that are compartmentalised into stocks and flows (see Fig. 2.1). The first set of relations, the social metabolism, constitutes biophysical stocks (human and livestock, artefacts) and energy and material flows. The quantities, type and weight of these flows are intricately linked to the system's population size and level of technology. In other words, the size of flows depends on the size of biophysical stocks of a social system. Territory (stock) and the appropriation of net primary production (flow) comprise the second set of relations, the colonization concept. The third set of relations, namely functional time use, entails human population in its stock category, and reproduction rates, life time/labour time in its corresponding flow category. All three sets of relations are interrelated with each other. Population size and demographic trends, for instance, determine the need for extraction, production and consumption of material, energy and land resources. At the same time though, a limited biophysical resource base can only sustain low human and livestock population numbers. The amount of human population is also decisive for a system's availability of human time resources. Having said this, however, the availability of human labour time does not depend on population size alone, but also on a system's age structures and culturally defined sexual labour division. The rural economist Ester Boserup (1965) has been instrumental in showing the interconnectedness between the availability of land, population size and labour time in agrarian societies, asserting that demographic pressure invariably favours more intensive land use practices which, in turn, require additional labour time investments.

1.4 Fieldwork Among the Tsimane'

When I had first joined a seminars cycle at the Institute of Social Ecology in Vienna, on rather non-committal grounds, I certainly did not know where this adventurous journey would take me. Admittedly, I had become intrigued by the novel biophysical approaches of quantifying material and energy flows between societies and nature as a way to measure (un)sustainability. To a young student like me, who was interested in learning about sustainable development, it seemed a perfect start to a more profound apprenticeship. In light of the up till then limited array of empirical results obtained from sociometabolic field studies, my spontaneous proposal to conduct a local biophysical study of a still fairly traditional social system was wholeheartedly encouraged by Prof. Fischer-Kowalksi and she accepted to supervise my thesis in early 2004. My undertaking would represent the third research endeavour (see Grünbühel et al. 1999; Grünbühel et al. 2003; Mayrhofer-Grünbühel 2004; Singh 2003) of a local setting 'at the periphery', a system, within which certain features would resemble those of historical agrarian regimes. Local approaches though were still in their infancy and I was warned of the thorny methodological problems I would come across in the field; it was as if, so I sensed at first, questioning the usefulness of the approach altogether. Yet, at least from my side, the possibility to take leave, at least for some time, of my daily routine and to discover some interesting, new facts filled me with immediate excitement and growing anticipation. As I thus set in train my descent into the revealing and yet unknown depths of empiricism, a whole fascinating journey still lay before me.

When it came to deciding on the fieldwork location, my choice of the geographical setting was clear from the outset: the Amazon Basin. As a region of the globe that has spurred the curiosity of natural and social scientists alike, it had also captured my imagination for a long time. To me, the Amazon evoked vividly strong images of wild forest vastness, verdure and somewhat complete peacefulness. At the same time though, I was familiar with mounting scientific literature sources all stating the dangers of accelerated deforestation rates and the loss of living space for indigenous communities. The call for a concrete sustainable development initiative, or at least a starting point, was becoming ever more pronounced. In retrospect, I probably hoped to contribute with some novel, albeit modest, local level insights to this inherently vast enterprise of sustainable development. Admittedly, I also liked the exoticness – at least for my fellow student colleagues of sociology – of my research. There was yet another, less personal, justification for this geographical choice. All previous biophysical case studies were conducted in Southeast Asia, namely in Thailand, Laos and the Nicobar Islands; a different case would possibly contribute with new revealing information. The choice of my study area in the Amazonian tropics therefore just seemed the perfect solution.

After a period of thorough research and experiencing the difficulty of trying to establish contacts from a distance, my choice fell on an indigenous community in the Bolivian Amazon: the Tsimane', a group of foraging horticulturalists inhabiting the Amazonian plains along the Rio Maniqui, a distant tributary to the Amazon. Despite the relative scarcity of historical documents, the Tsimane' are no new arrivals on the ethnographic scene. I had enjoyed reading most of the early intriguing ethnographies about the Tsimane', among which the works of Nordenskiöld (2001[1913]), Metraux (1948) and Pérez Diez (1983) figure most prominently. They were generally quite descriptive and comprehensive and I could browse across a variety of subjects ranging from religion to the economic and socio-cultural sphere. More recent research efforts though have moved towards writing more theoretically focused monographs, unfolding a line of arguments supported by empirical records. This latter category includes the thesis work of various young doctoral students. While Daillant (1994) and Ellis (1996) both concentrated their efforts on social relations and patterns of kinship, Piland (1991) gave a notable description of agricultural practices and indigenous soil management. Chicchón's (1992) main concern revolved around the impacts of market involvement onto Tsimane' subsistence activities. Byron (2003) wrote about the effects of market integration on health, while Reyes-García's (2001) work analyses the ethno-botanical knowledge of the Tsimane'. My research coincided with the coming together of a wide-reaching research programme named TAPS,[2] an initiative coordinated by Brandeis University in Massachusetts, USA. Through this network, American students of anthropology are given the possibility to do empirical research among the Tsimane'.

Preliminary investigation revealed that a number of Tsimane' communities were beginning to undergo certain socio-economic transformations, sparked by frequent contacts with logging firms, missionaries or other socio-political influences. Most of them, however, still maintained the most salient characteristics of a traditional way of life, fast vanishing in other dialect groups in the area. Likewise, Tsimane' families mostly live in community settings, with fairly clear territorial divisions between them. The fact that there would generally be a no-man's land in terms of settlement areas (though within the official Tsimane' territorial boundaries) of a couple of hours up to a day's walk or canoe trip between the villages, would facilitate the task of demarcating system boundaries. This, coupled with the fact that the Tsimane' still go about their interaction with nature in a primarily domestic setting,

[2]For more information on TAPS: http://people.brandeis.edu.

added another two points in favour of my choice. Also, I had established contacts with former researchers as well as a number of institutions which all offered their support. My remaining time until departure could thus be divided between general planning, organising and drawing up of survey material. It was Johnson's (2003) work on the indigenous Matsigenka people and Descola's (1996) monograph on the environmental relations of the indigenous Achuar ethnic group that both greeted me on and guided me through the threshold of my empirical world. Full of determination and probably a little trepidation, it was in this scholarly atmosphere that I left for the Bolivian Amazon in August 2004. The challenges and opportunities of fieldwork in a remote indigenous community far from our pace of life lay ahead.

Equipped with a measuring tape, piles of papers and a formal letter of introduction from my university, I arrived in San Borja, a typical lowland town that owes its existence to the thriving cattle farming business. A bustling commercial hub during the daytime hours, recurring power cuts especially during the evening hours and the complete lack of electricity along the outer reaches of the town, can make the nights seemingly long and dark. It hosts the seat of the municipal government that is responsible for the administration of decentralised development programmes in the area. Most shops and eating places are lined up along the market area, where all kinds of market commodities, from mosquito nets to bicycles, can be obtained. In essence, San Borja acts as a magnet for the Tsimane' to spend a day away from their community; they are usually seen around the lively market place, where they sell or barter their agricultural produce in exchange for foods, clothes or other trade articles. At night though, most Tsimane' return to their hamlets. Among the locals I was generally met with surprise on learning of my intention to live with the Tsimane'. Despite the fact that some appeared more sympathetic than others and praised the quality of Tsimane' produce, I did sense a generally prevailing hostile attitude towards the Tsimane' community. Despite the fact that they live right on the edge of the Tsimane' territory, the common image projected onto them was frequently that of an uncultured tribal people that have absolutely no interest in regional integration. It seems though that this experience was no unique incident. In 1972, Johnson (2003: 4, 5) had a similar experience on his way to fieldwork with the indigenous Matsigenka in the Peruvian Amazon. Upon articulating his upcoming fieldwork among the Matsigenka in a restaurant in the market town of Quillabamba, located just a stone's throw away from the Matsigenka territory, he was warned of being shot full of arrows by those savage tribes. Allegedly near in terms of geographical distance, it seemed that I was still quite far from my final destination.

Hearing about the novelty of a foreign student arriving, I was fortunate enough to be welcomed by a small group of Bolivian and American anthropologists who had their research base in San Borja. Owing to their support, I was facilitated contact with the *Gran Consejo Tsimane'* (Tsimane' Council), the political authority of the Tsimane' ethnicity, who soon granted me permission to work in their territory and supported me in the selection of a suitable community to work and reside for the months to come. In line with my research objective, I was generally looking for a representative place with a still relatively high subsistence economy and some (outside) opportunities for cash income. Likewise, the number of households should be manageable for a sole researcher. Equally important, I thought that doing research in a community within the wider limits of the Beni Biosphere Reserve, a biosphere reserve of great ecological importance, would give me the opportunity to observe the effect of conservation policy on actual resource use strategies. Relative proximity to San Borja would be an additional add-on, since I would be able to return once in a while for the comfort of a shower and for, probably more pressing, mosquito repellent. Yet what really counted in the end would be the readiness of the villagers themselves to accept a stranger living among them. The community of Campo Bello seemed the perfect place to be. Not merely intrigued by its euphonious name which literally translates into 'beautiful field', it embodied all the features I was looking for. Adding to this, even a telephone would be at my disposal, I was told by the head of the Tsimane' Council. Surprised at the idea of finding a telephone in a place where people are not even connected to electricity, I would soon find the telephone to be a remnant of a past development effort. The community of Campo Bello, an approximate 4-h walk from San Borja, locates along the Rio Maniqui and comprised 41 households at the time of research in 2004. Situated near a gallery forest and savannah lands, it is in fact located within the buffer zone of the biosphere reserve. People engage in subsistence agriculture and sell their agricultural surplus and forest items to itinerant traders or directly in San Borja.

In the early hours of a clear early September morning, Everisto from the Tsimane' Council and I finally took off on his motorbike to Campo Bello. Leaving behind the urban infrastructure of San Borja we followed a dirt road that slowly turned into a continuously more rugged and narrower forest path as we reached the banks of the Rio Maniqui. Upon reaching the first Tsimane' dwelling in San Antonio, a neighbouring community of Campo Bello, we were greeted with a friendly *Najjoi'* (good morning). Upon catching sight of us, passers-by would exclaim '*Hana mura mi*', where are you going? It is an expression that I learnt quickly, as it was of common usage upon any sort of casual encounter outside the world of domestic space. We managed to move forward slowly on the back of the motorbike until the narrow forest trail suddenly gave way to the sprawling river. Standing at the waterfront, I could simply not imagine how crossing the river would ever become possible. Yet after placing the motorbike, my baggage and ourselves into a small-sized wooden canoe, a Tsimane' boy about the age of eight skilfully paddled his way through the fairly turbulent river waters. It was my first experience of the skilful verve the Tsimane' generally exhibit when interacting with their natural environment. After reaching the other river bank, it was still a half-hour walk before reaching the community house and the school premises, both of which represent the village centre. Since we arrived during the morning break, children were playing football on the pitch outside the school building. Although some noticed our presence, their general fervour for the game remained largely undisturbed. Behind the school in the community house, an eight-legged traditional wooden construction with a palm-thatched roof, a couple of people had gathered, seemingly expecting the stranger (I later found that the Tsimane' Council had announced my visit via their daily radio programme). My introduction was given as a researcher who wanted to stay among the community members for the 6 months to come. After explaining the broad goals of my research, the people present, though seemingly rather indifferent to my presence, gave their consent to participate in the research. The bilingual teacher, known as the *profe*, pointed to an old kitchen structure, a four-legged wooden house without any walls but a thick palm roof, where I could fasten my hammock. Located in the immediate vicinity of the school, the building had long ceased to function as a kitchen, while its rooftop occasionally served as a resting place for poultry. Soon after all was settled, Everisto embarked on his return to San Borja.

During the first few days I was occupied with familiarising myself with the local environment. I took long walks, organised my research papers and thoroughly studied the language basics with the help of the didactic material I was given by my anthropological friends in San Borja. As I was basically 'hanging out', I remembered the entertainingly written text by de Munck (1998: 41–45) who classified the period of hanging out by anthropologists in three distinct stages. The first was referred to as the 'stranger' stage, a process characterised by trust building and social learning, in which I found myself in over the first few weeks. One day, when the teacher and his wife offered me to join a few classes, I leapt at the opportunity to become more immersed in the local village patterns. I suppose that this was the moment when I entered the 'acquaintance' stage. The children were curious about my note-taking and would sometimes join me on my wanders around the village or guide my first steps into the forest. In retrospect, I was quite happy that a biophysical analysis requires such a great deal of observation, measurements and estimations that can largely be done 'from a distance' without too much interference in the peoples' lives. Likewise, my continuous and clearly odd queries caused giggling and laughter to most of my young Tsimane' companions.

While a few men are bilingual or may at least have some notion of Spanish, most people and especially women rarely communicate in Spanish. For the first rounds of interviews, therefore, I was offered translation services by Roman, the local village leader. On his third day though, Roman got bored of my unrelenting questioning and left me to my own devices. I still remember a certain frustration which I confided in my field diary: 'people here are indifferent to me, but friendly. Roman got bored, he wants to work on his field, he says. We are still missing half of the households, some have left and nobody knows when they are back. I have no choice but wait' (fieldnotes 10/09/04). I had to get used to a different pace of life, take things as they came along. Nonetheless, it was not until the teacher's wife Juanita offered me her help in installing a raised platform bed for me, things started to change. To make my stay more homely, I was given small pieces of self-made furniture, some women provided me with floor mats or woven bags and Roman and his wife showed me how to construct a wall out of local materials. Thankful for their generous and tolerant gestures, I tried to return my gratitude in various ways[3] and noticed that the initial barrier between the people and I gradually diminished.

[3]I would, for example, buy medicines, clothes or foodstuffs on my monthly trips to San Borja.

For the remaining survey rounds, some local boys took turns in assisting my research. Looking back, my fondest memory is of sitting outside the house in the early-morning mist awaiting my Tsimane' companions in pleasant anticipation of the adventures – an unpretentious forest stroll, a fishing trip or honey gathering – of that day. I suppose that these moments were the closest I came to the final, or, in de Munck's term, 'intimate' stage of participant observation.

With piles of interview records, field notes and myriad photographic material I returned home in late February 2005, with the interpretation and analysis of the empirical data still lying ahead. On my second visit from April to May 2006, things were altogether much more familiar. The time of my stay coincided with the annual rice harvest period, a time when production, consumption and time use patterns are slightly different from other periods throughout the year. Also, my second visit intended to add some additional insights into the dynamic changes of biophysical processes.

1.5 What the Book Says

In the remainder of this chapter I present an overview of the book, which grapples with the immense challenge of moving towards (and not away from) a more sustainable future. This book has been inspired by a larger ongoing research programme in cooperation between the Institute of Social Ecology in Vienna and its partners worldwide, whose individual research endeavours are all driven by the same overall assumption: a transition to a more sustainable sociometabolic regime calls for fundamental transformations – such profound changes as did the transition from agrarian to the industrial regime, a transition that is still ongoing in many parts of the world (Fischer-Kowalski and Haberl 2007: 19-20). To this end, enhancing our understanding of what such a sociometabolic transition entails in terms of the biophysical structures of a social system and its impacts on nature provides a crucial starting point. In such a light, the book's motives are driven by two overlapping interests: a generic and a comparative approach to sociometabolic regime transitions. As concerns the former, the study aims to provide a detailed description of Campo Bello's metabolic profile, thereby depicting the specific pathways and bottlenecks for future developments. As the concerns of the first motivation diminish, so those of the second increase; there is no clear dividing line. No less important is the second interest of this book that is to attain an in-depth understanding of patterns found by comparing the biophysical features of different agrarian sociometabolic regimes across the so-called 'developing world'.

The book is organised into six forthcoming chapters with the first four chapters largely focussing on the local dynamics of transitions. The monograph opens by introducing the scientific research framework and moves on to guide the reader from the historical to the contemporary field setting. Departing from the grounds of hermeneutics, quantifying the use of materials, energy, land and time then provide the empirical heart of the biophysical local study. The results of the second more complex and far-reaching motivation of this book, which begins to develop mainly in Chapter 6, are admittedly more humble.

Chapter 2 provides the theoretical foundation. It opens with an introduction of the concept of sociometabolic transitions, the notion that provides the heart of the transition study paradigm within which my study is embedded. What follows is the presentation of the MEFA framework, the central device for analysing and describing the metabolic exchange relations between societies and their natural environment. It combines the three conceptual tools of 'social metabolism', 'colonization of terrestrial ecosystem' and 'life time-labour time'. In chronological order, the chapter continues to scrutinise the concept of social metabolism, starting with a historical overview of the roots of this concept, which has its origins in a variety of academic disciplines. What then follows is a description of material and energy accounting methods, the specific tools for quantifying the metabolic exchange relations between humans and their natural environment. The focus then shifts to discussing a complementary second set of relations, the colonization concept, which entails a discussion on territory and the impact of land use. To this end, the calculation of the 'human appropriation of net primary production', or simply HANPP, will be introduced as an example of measuring the impact of human intervention upon the ecosystem. The chapter concludes by attempting to draw out a further, undeniably less developed, set of relations; the notion of 'life time-labour time' constitutes the third conceptual base upon which (un)sustainability indicators can be elaborated.

Chapter 3 introduces the reader to the ethnographic setting of the Tsimane'. In line with the important dimensions that shape a society's relations with nature, I have distinguished four different sections. The first part entails a discussion on the Tsimane' territorial space, listing territorial features and the biogeographical conditions of the area the people inhabit. The territorial concept has been given broad attention in biophysical studies, as defining system boundaries in many empirical settings presents a real challenge. The second suitably titled 'landscape and cosmos' domain bridges the cultural with the natural sphere and gives a detailed account of the cultural features that direct the Tsimane' appropriation of nature by means of farming, foraging and fishing activities. The third section provides a brief historic overview with a particular focus on early sporadic encounters with outsiders to outlining the reasons for establishing more permanent links with the regional economy. Finally, the chapter draws to a close with a discussion on the Tsimane' social world.

Chapter 4 then zooms into the local world of Campo Bello, a Tsimane' village that is located along the meandering course of the lower Rio Maniqui. What is this agrarian sociometabolic regime like in terms of both its biophysical structures and its transformative impact on nature? A general village description and a portrayal of changing population dynamics set the stage. What follows is a comprehensive biophysical analysis, using the standardised MEFA framework methodology. It starts with tracing the material and energy stocks and flows (in terms of quantity and weight) that enter the community, get transformed and depart again. Environmental constraints and pressures, which may occur at any of these 'stations' throughout this socio-economic metabolic process, will be identified. The empirical analysis of the colonization concept will bring to light the impact of the peoples' specific land use practices. As agriculture provides the main socio-economic activity for the people of Campo Bello, a lengthy description of the local agricultural cycle and the human labour involved precedes the empirical analysis of the HANPP approach to measure the human impact on the local ecosystem. Finally, the insights generated are woven together into a biophysical discussion that, in light of sustainability transitions, highlights possible and impossible futures for the local community of Campo Bello.

Chapter 5 expands the biophysical discussion to include yet another resource: human time. The point of departure for this chapter is the assumption that for traditional societies the extent of working time is ultimately responsible for the metabolism of a society. Since traditional societies merely invest human labour, the amount of human labour time is directly linked to agricultural production. For that reason, this chapter examines the productivity of time in the local system. How much time can be supplied by different age–sex categories and to what degree are these resources invested to reproduce the social system? Drawing on the systems theory by Luhmann (1984, 1995), I will present the possibilities and constraints of the various social subsystems: the person system, the household system, the economic system and the community system. Along these lines, three questions will be addressed. First, how much labour time can the local system provide? Second, who contributes what time resources to the reproduction of the four functional subsystems? Third, a final section looks at the dynamics of changes in human time use. And what strategies are being followed by the local people to increase the productivity of their time?

Chapter 6 is concerned with the second motivation of this book: obtaining a better understanding of global transition patterns through a comparative approach. How 'typical' is the village of Campo Bello in terms of its biophysical features? Does it show any similarities with other local cases and, if so, how can they contribute to the global sustainability transition debate? Undoubtedly, the aim of this chapter is more modest than these questions would require for a fully comprehensive answer. Without longitudinal data (and/or formal models) it is hard to demonstrate transition dynamics. What I present is merely a comparative biophysical assessment of four local case studies, all conducted in different biogeographical settings and historical world contexts. What they have in common, however, is the fact that they are all located 'at the rural periphery' of developing countries, where, at different scale levels and with different degrees of interaction, transition processes towards industrialisation are well under way.

Chapter 7 seeks to knit together those strands of insights assembled throughout the course of the book. It presents a selective summary of what we have learnt from this explorative journey and critically discusses how these new insights may apply to a broader context. Equally important, this chapter also discusses some methodological shortcomings and constraints in order to attain a full cognition of consumption patterns, cultural practices and social institutions.

A book of this kind is neither intended nor able to be exhaustive, and I have tried to strive for coherence rather than present an unwieldy synthesis. The coherent frame was made possible by relying on an already existing scientific state-of-the-art approach, the MEFA framework. I could certainly take advantage of this standardised methodology and cross-check my biophysical results with previously conducted local research to which, undoubtedly, the merit goes (see Grünbühel et al. 2003; Mayrhofer-Grünbühel 2004; Singh 2003; Singh and Grünbühel 2003). I believe that because of the availability of this consistent framework I was able to further advance some specific measurements and estimation methods. Any other assertion would certainly be adorning oneself with borrowed plumes. Yet this book aims to go beyond simply adding more empirical data to the array of local biophysical studies, as it also intends to advance the theoretical framework by adding the use of time to the sociometabolic concept we are dealing with. A kind review by my (then) 'doctoral mother' Marina Fischer-Kowalski once portrayed my work as a useful and creative guidance for much-needed future empirical local systems research, on the one hand. On the other hand, her evaluation commends my analysis of local time use which in her view 'provides a framework of system reproduction that allows a macroscopic glance at human time as a resource and constraint of societies'. In such a light perhaps, this book might best be regarded as a contribution towards making the MEFA concept more operational for its application to contemporary local systems in transition. As concerns my attempts to empirically integrate the use of time, the findings shall be seen as a 'work in progress' with further methodological refinement still pending. The stance this book takes, however, is that changing working times may, in the shorter or longer run, put a burden on the whole society or individual segments (e.g., women or children) of the society in transition.

References

Boserup, E. (1965). *The conditions of agricultural growth: The economics of agrarian change under population pressure*. Chicago: Aldine/Earthscan.

Byron, E. M. (2003). *Market integration and health: The impact of markets and acculturation of the self-perceived morbidity, diet, and nutritional status of the Tsimane' Amerindians of lowland Bolivia*. PhD thesis, University of Florida.

Chicchón, A. (1992). *Chimane resource use and market involvement in the Beni biosphere reserve, Bolivia*. PhD thesis, University of Florida.

Cudworth, E. (2003). *Environment and society*. London: Routledge.

Daillant, I. (1994). *Sens dessus dessous: Organisation sociale et spatiale des Chimane d'Amazonie bolivienne*. PhD thesis, Laboratoire d' Ethnologie et de Sociologie Comparative, University of Paris.

de Munck, V. C. (1998). Participant observation: A thick explanation of conflict in a Sri Lankan village. In V. C. de Munck & E. J. Sobo (Eds.), *Using methods in the field: A practical introduction and casebook* (pp. 39–56). USA: Altamira Press.

Descola, P. (1996). *In the society of nature: A native ecology in Amazonia*. Cambridge: Cambridge University Press.

Ellis, R. (1996). *A taste for movement: An exploration of the social ethics of the Tsimanes of lowland Bolivia*. PhD thesis, University of St. Andrews.

Eurostat. (2007). *Economy-wide material flow accounting. A compilation guide*. Luxembourg: European Statistical Office.

Fischer-Kowalski, M., & Haberl, H. (2007). Conceptualizing, observing and comparing socioecological transitions. In M. Fischer-Kowalski, H. Haberl (Eds.), *Socioecological transitions and global change: Trajectories of social metabolism and land use* (pp. 13–62). Cheltenham, UK, Northampton, USA: Edward Elgar.

Fischer-Kowalski, M., & Weisz, H. (1999). Society as hybrid between material and symbolic realms. Toward a theoretical framework of society–nature interrelation. *Human Ecology, 8*, 215–125.

Fischer-Kowalski, M., & Weisz, H. (2005). Society as hybrid between material and symbolic realms. Toward a theoretical framework of society–nature interaction. In M. Redclift, & G. Woodgate (Eds.), *New developments in environmental sociology* (pp. 113–449). Cheltenham and Northampton: Edward Elgar.

Godelier, M. (1986). *The mental and the material: Thought economy and society*. London: Blackwell Verso.

Grünbühel, C. M., Schandl, H., & Winiwarter, V. (1999). *Agrarische Produktion als Interaktion von Natur und Gesellschaft: Fallstudie SangSaeng*. Vienna: Social Ecology Working Paper 55.

Grünbühel, C. M., Haberl, H., Schandl, H., & Winiwarter, V. (2003). Socio-economic metabolism and colonization of natural processes in SangSaeng village: Material and energy flows, land use, and cultural change in northeast Thailand. *Human Ecology, 31*(1), 53–87.

Johnson, A. (2003). *Families of the forest: The Matsigenka Indians of the Peruvian Amazon*. Berkley: University of California Press.

Luhmann, N. (1984). *Soziale Systeme: Grundriss einer allgemeinen Theorie*. Hamburg: Suhrkamp Taschenbuch.

Luhmann, N. (1995). Sysatem and function. T. Lenoir, & H.U. Gumbrecht (Eds.), *Social systems*. (pp. 12–58). Stanford: Stanford University Press.

Mayrhofer-Grünbühel, C. (2004). *Resource use systems of rural smallholders. An analysis of two Lao communities*. PhD thesis, University of Vienna.

Metraux, A. (1948). The Yuracare, Mosetene and Chimane. In J. Steward (Ed.), *Handbook of South American Indians*. Washington: Smithsonian Institution, Bureau of American Ethnology.

Moran, E. F. (2006). *People and nature: An introduction to human ecological relations*. Oxford: Blackwell Publishing.

Nordenskiöld, E. (2001). *[1913] Exploraciones y aventuras en Sudamérica*. La Paz: APCOB.

Parris, T. M., & Kates, R. W. (2003). Characterizing a sustainability transition: Goals, targets, trends and driving forces. Retrieved April 10, 2009, from http://www.pnas.org/cgi/doi

Pérez Diez, A. (1983). *Etnografía de los Chimane del oriente boliviano*. PhD thesis, University of Buenos Aires.

Piland, R. A. (1991). *Traditional Chimane agriculture and its relation to soils of the Beni biosphere reserve*. Bolivia: Master thesis, University of Florida.

Raskin, P., Banuri, T., Gallopín, G., Gutman, P., Hammond, A., Kates, R. W., & Swart, R. (2007). Great transition: Where are we. In great transition: The promise and lure of the times ahead. Retrieved May 12, 2009, from http://www.eoearth.org/article/Great_Transition:Where_Are_We

Reyes-García, V. (2001). *Indigenous people, ethno-botanical knowledge and market economy: A case-study of the Tsimane' Amerindians in lowland Bolivia*. PhD thesis, University of Florida.

Schandl, H., & Grünbühel, C. (Guest Eds.). (2005). Southeast Asia in transition. *International Journal of Global Environmental Issues*, *5*(3/4), 115–118.

Schellnhuber, H. (1999). "Earth System" analysis and the second Copernican revolution. *Nature, 402*, C19–C23.

Singh, S. J. (2003). *In the sea of influence: A world system perspective of the Nicobar islands*. Lund: Lund University.

Singh, S. J. (2006). *The local base of transitions in developing countries*. Vienna: Presentation at the seminar at the Institute of Social Ecology.

Singh, S. J., & Grünbühel, C. M. (2003). Environmental relations and biophysical transition: The case of Trinket island. *Geografiska Annaler*, *85B*(4), 187–204 (distributed by Blackwell: UK & USA).

Wackernagel, M., & Rees, W. (1996). *Our ecological footprint, reducing human impact on the earth*. Gabriola Island, Philadelphia: New Society Publishers.

Chapter 2
Sociometabolic Transitions and the MEFA Toolkit: Concepts and Methods

Abstract This chapter provides the theoretical framework upon which the case study is based. It first introduces the concept of sociometabolic regime transitions and then moves on to examine the MEFA framework, the core instrument for describing the metabolic exchange relations between a society and its natural environment. Using this instrument helps us to assess the environmental costs (caused by flows of energy, material and a certain type of land use) as well as the social costs (through labour time). To this end, the MEFA framework comprises three interrelated sets of metabolic relations, each of which will be discussed separately. 'Social metabolism' constitutes the first set of relations, followed by 'colonization' as the second set and 'functional time use' as the third set of metabolic relations.

2.1 Introduction

The main goal of sustainability science research is to unite valuable efforts across the natural and social sciences in order to find sustainable pathways toward a common global future. While approaches may differ with regard to the disciplines involved, there exists, however, common agreement that human intervention is indeed an indispensable task to ensure sustainability. Only through deliberate human agency can local and global development paths be steered toward a state that is superior to the industrial regime we currently live in, or many social systems – in fact more than half of the world's rural population – are currently heading for. Yet to act accordingly, we first of all require a baseline of the current situation. One such baseline approach, namely a description of the biophysical exchange relations between a society and its environment, is presented by this book, and this chapter deals with the theoretical framework that is needed to do so. To this end, the theory of sociometabolic regime transitions will be presented, which somewhat provides the theoretical

L. Ringhofer, *Fishing, Foraging and Farming in the Bolivian Amazon: On a Local Society in Transition*, DOI 10.1007/978-90-481-3487-8_2,

umbrella for the empirical biophysical study that follows. I will then proceed to give a detailed account of the MEFA framework, the core instrument for analysing the metabolic exchange relations between human societies and their natural environment. In so doing, the MEFA framework measures both, the pressures on the environment (caused by constant flows of energy, material and a certain type of land use) as well as the burden on people in terms of working time (through changes in time use). In this respect, I refer to the framework as a conceptual and methodological toolkit, since it contains three interrelated sets of metabolic relations: (1) social metabolism, (2) colonization of terrestrial ecosystems and (3) functional time use. Each concept will be discussed individually, together with their respective methodological tools for empirical application.

2.2 Sociometabolic Sustainability Transitions

2.2.1 Characterising a Sustainability Transition

Sustainability has occupied a place on the global agenda since at least the Brundtland Commission's report *Our Common Future* in 1987 (Brundtland Report 1987). With ever mounting global environmental concerns such as water and biodiversity, its place has become ever more prominent in the 1990s. European and US approaches to finding pathways toward global sustainability have been outlined in a variety of publications. European thinking, for example, involves Schellnhuber and Wenzel's (1998) research on earth systems analysis or the European Commission's Fifth Framework Programme (1998); among the US publication it is the United States National Research Council, Board on Sustainable Development (1999) *Our Common Journey: A Transition toward Sustainability* which figures most prominently.

With the advent of the twenty-first century, the field of sustainability science was created. It was a long overdue attempt to merge views, practices and perspectives from different disciplines across the natural and the social sciences. Yet despite a certain level of vagueness as to the methodological approaches among this emerging scientific community, something new could be felt 'in the air'. Clark and Dickson (2003), for example, voiced their general sense of excitement about this new research programme, calling it 'rather a vibrant arena...its scope of core questions, criteria for quality control, and membership are consequently in substantial flux and may be expected to remain so for some time'. While methods and approaches may differ among the interdisciplinary research community, the core of the research interest is unalterably shared across the scientific landscape: the focus on the dynamic interactions between natural and social systems, with equal attention to how social change shapes the environment and how environmental change at the same time influences society. Some of the key researchers in this field have been looking into long-term trends in environment and development interventions (see, for example, Kates and Parris 2003), examined ecosystem-specific resilience or vulnerability levels of nature–society systems (see, for example, Turner et al. 2003) or focussed on modelling the dynamic interactions between nature and society (Schellnhuber 2003).

Parris and Kates (2003) have been instrumental in characterising a sustainability transition. A sustainability transition as described by the Board on Sustainable Development of the National Research Council is seen as one where a stabilising world population (1) meets its needs and reduces hunger and poverty while (2) maintaining the planet's life support systems and living resources. While the first of these goals is much more institutionalised in specialised agencies and multilateral programmes (e.g., the United Nations Development Programme, the United Nations Educational, Scientific and Cultural Organization), there is generally little consensus as to the second goal. With the exception of the Montreal or Kyoto Protocols, most environmental agreements lack the global character so inherently needed for a worldwide sustainability transition to take place. Juma (2003) makes the point that a Sustainability Transition requires a global process of social learning involving profound changes in the way human beings interact with the natural environment as well as with each other. This global process, however, is still largely in its infancy.

So far, what the scientific community agrees on are the general features characterising a transition. First, a sustainable future development innately calls for active and deliberate intervention of human agency in existing system dynamics. It is no longer believed that transitions occur according to endogenous, system-imminent dynamics that dictate the direction of development. This is indeed in stark contrast to prevalent development approaches propagated by a number of actors – the World Bank for example – promoting, in light of progress, industrial market integration reforms in an effort to 'develop' rural subsistence communities. Notwithstanding, today's discourse on sustainability somewhat contradicts this concept of progress. With a sustainable society as considered something 'superior to the current state', a transition to a more sustainable global society should be 'more a matter of reason than of passions' and not necessarily considered 'the logical and inevitable next stage' (Fischer-Kowalski and Haberl 2007: 7). Secondly, another unique feature of the transition notion is its spontaneous character. A transition may occur unprompted; once initiated, however, it has the ability to remain in a state of dynamic equilibrium for some time. Following this train of thought, a transition leads to a completely new condition; we may all feel this transition taking place around us, but are unable to predict its exact outcome. The main responsibility of sustainability scientists therefore is to pick up on past, present and future trends and their driving forces in order to comprehend how these are actually reshaping society–nature interactions at both, the local and the global level.

2.2.2 Sociometabolic Regimes and Transitions: Theoretical Considerations

The theory of sociometabolic regime transitions is one such endeavour to capture these long-term trends needed to understand the dynamics at the society–nature interface. Covering large time frames, from decades to even centuries, the specific research focus centres on structural change of the inter-linkage between social systems (e.g., local community, region, state) and natural systems (see Fischer-Kowalski and Haberl 2007). Sociometabolic regimes, in this context, are characterised by typical patterns of biophysical interaction between the social and the natural system.

For the theoretical concept to become applicable, a number of conceptual idiosyncrasies call for clarification. First is the definition of the social system, the core point of reference for our sociometabolic analysis. For this purpose, the social system (human society) is defined as a hybrid, a structural coupling of both realms, the cultural system with biophysical elements (for more detail, see Fischer-Kowalski and Weisz 1999).[1] Our understanding rests on the notion that social systems need to reproduce themselves culturally (via communication, see Luhmann 1984, 1995) as well as biophysically in order to guarantee continuous communication flows. As to the cultural sphere, Luhmann's system theoretical make-up has been elementary for two reasons: it encourages thinking in terms of functional differences and the production of boundaries in relation to outside elements (self-referential systems). At the same time, his systemic approach facilitates functional differentiation of the internal complexities governing each of systems we observe. But Luhmann's arena of influence stops sharply at the threshold of the second realm: the biophysical sphere. In his view, biophysical structures indubitably belong to the social system's environment and hence are directed by a different set of rules. The sociometabolic understanding is different though, as biophysical structures are considered as an equally important component of the social system. The core idea of the sociometabolic theory is in fact the following: the cultural system interacts with the natural system via these biophysical structures; and in so doing, each system follows its own system-specific logic.

When a society interacts with its environment, it does so by the (often unintended) exchange of material and energy and (intentionally) by means of applying certain technologies and labour in order to increase the utility of the natural system for the society. This in turn generates impacts in the environment to which societies then have to respond. As the social system is structurally coupled with its environment, both systems mutually depend on, influence and equally constrain each other. In this respect, certain patterns of society–nature interactions that retain similar

[1] This is actually in stark contrast with more common definitions of society found in sociology that tend to stress the collective importance for a specific group of people that shares certain commonalities (see, for example, Giddens 2002; Jenkins 2002). Lenski (1974) takes the term to a more evolutionary level as he differentiates between different societies according to their level of technology and economy, for which he produced a system of classification based on social inequality and the role of the state.

features over longer time-spans including transition periods – referred to as sociometabolic regimes – can be observed (Fischer-Kowalski and Haberl 2007: 14). Historically, three different sociometabolic regime patterns can be distinguished: the hunter and gatherer regime, the agrarian regime and the industrial regime. Each has a specific set of basic biophysical characteristics such as resource use levels, the use of labour time, demographic growth rates and settlement patterns. In other words, each of these regimes is characterised by a certain metabolic profile that relates to a certain set of impacts upon the environment.

Sieferle (1989, 2001, 2003) is considered the founding father of the theory of sociometabolic regimes and others (Fischer-Kowalski et al. 1997; Sieferle et al. 2006) have since further elaborated and refined certain assumptions. In his publication *Sustainability in a World History Perspective*, Sieferle (2003) provides an appealing historical account of sociometablic regimes in world history, claiming the availability of energy to represent the major element for distinguishing socioecological systems. In his view, socioecological systems that share the same source of energy will equally display a similar set of other biophysical features that are largely independent of biogeographical conditions. Along the same lines, he discusses the sustainability of previous regimes, pointing to the specific sustainability threats of each regime pattern (see Fischer-Kowalski and Haberl 2007; Krausmann et al. 2003; Krausmann et al. 2008). Hunters and gatherers, to start with, do not deliberately intervene in transforming plant biomass (e.g., through farming techniques); their regime is therefore based on an uncontrolled solar energy system. As hunters and gatherers must more or less live on the resource base they have available in their territory, the only sustainability threat they pose is in the form of overexploitation of natural key resources. Agrarian societies, on the other hand, can be characterised by a regime of active or controlled solar energy utilisation through the intervention in nature by means of biotechnologies (forest clearance and creation of agro-ecosystems) and mechanical devices (e.g., wind and watermills). The most pressing sustainability threat under this regime is sustaining high population growth rates while simultaneously maintaining the long-term productivity of agro-ecosystems (soil fertility). Moreover, these regimes most often are exposed to technological and political dependencies and have to deal with fluctuations of natural systems. The presently dominant industrial regime, dating back some three centuries, is based upon the exploitation of fossil fuels. Its sustainability is threatened by the limitations of its energy base, on the one hand, and the transformations it prompts in various natural systems. Its regime character, however, remains somewhat contested, since it relies on non-renewable material and energy resources. Its long-term existence is thus threatened, as it rapidly eats away its own life-sustaining resource base. Change is therefore bound to happen. In this sense, it is high time for searching transition paths that in the short and longer term come at the lowest burden possible for the environment as well as for human societies (Fischer-Kowalski et al. forthcoming). The MEFA framework described below offers one such toolbox to empirically apply the sociometabolic regime theory to individual social systems. In so doing, it is hoped to enhance our understanding of the system's specific pattern of society–nature interaction in order to advise appropriate and sustainable ways forward (Fig. 2.1).

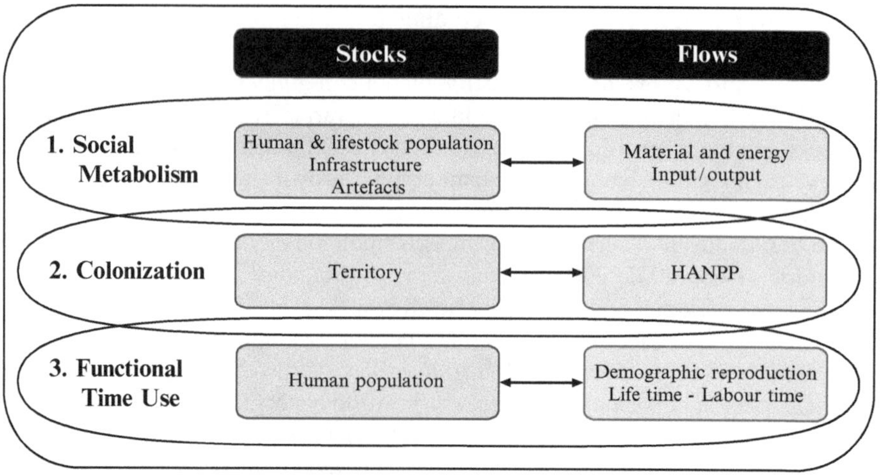

Fig. 2.1 MEFA framework with three sets of metabolic relations (adapted from Fischer-Kowalski and Haberl 2007)

2.3 The MEFA Framework Toolkit

The MEFA framework toolkit is the core instrument for analysing the energy and material exchanges between societies and their natural environment. If we apply the individual empirical tools we find in this methodological toolbox, we will obtain the following results (see Chapter 4): a detailed metabolic profile of the system under discussion, an analysis of the feedback loops shaping both, the social and the natural system, and a clear indication of the biophysical limitations the system is currently facing (in terms of material, energy, land and time constraints). What adds to the MEFA frameworks' attractiveness is its ability to establish an inherent link between biophysical indicators and socio-economic variables. It is through this ability that the MEFA framework has gained valuable recognition within the national accounting domain (Eurostat 2001, 2007). What the framework cannot deliver so far is the provision of a more dynamic, theoretical integration of all variables and their interrelation with each other. The model framework existing thus far contains conceptual tools and applicable methodologies (material flow accounting, energy flow accounting, human appropriation of net primary production, and functional time use) in order to derive at a sound metabolic profile of the society under discussion. This, in turn, sheds light on the possibilities and constraints the specific pattern of interaction between this society and its environment in fact entails.

The MEFA framework consists of three interrelated sets of metabolic relations (social metabolism, colonization of terrestrial ecosystems and functional time use), each of which I am going to describe individually over the next pages. I included a rather extended historical section on the intellectual origins of the term metabolism, pulling them together from different academic strands. For those only marginally interested in this lengthy historical account I suggest leapfrogging the next pages and resume reading at Section 2.3.1.2. Material and Energy Accounting; so as not to disrupt your reading flow.

2.3.1 Social Metabolism

The term metabolism finds its origins in biology and ecology. Yet the metabolic idea has been taken onto another level to explain the interaction of human societies with their natural environment: a society extracts primary resources and makes use of them as foodstuffs, machines, buildings, infrastructure, heating and other products until it finally returns them in the form of emissions and wastes (Fischer-Kowalski and Haberl 1998). Thus, metabolism as a social concept shall be somewhat perceived in a metaphorical manner, as internal biochemical conversions are replaced by external socially organised material and energy exchange processes between a society and its environment. In seeing society and its economy as an open system in constant physical exchange with natural systems, the notion of social metabolism implies a new conceptualisation of a society's pressure on the environment, on the one hand, and its dependency on other social systems, on the other hand (Schandl et al. 2002). Material and energy accounting are methodological tools for calculating socio-environmental pressure indicators which have become widely recognised standards for calculating socio-economic pressures upon the environment. Simultaneously, such an analysis further assesses in detail the society's use of resources and makes qualifications as to whom they benefit. In one of her earlier works, the social ecologist Marina Fischer-Kowalski (1997) enthusiastically referred to the concept of social metabolism as a 'rising conceptual star', in which she gives a remarkably detailed historical account of the roots and traces of this term across the social and natural disciplines. A great deal has happened since and today, more than a decade later, we have witnessed a lot of fine-tuning. But to reflect further here would be to end the voyage before it has even started. So I suggest, let us journey back in time and start by examining the birth of the metabolic concept in biology and how it has since been adopted and developed further by other disciplines.

2.3.1.1 Scholarly Origins of Social Metabolism

The academic history of the concept of metabolism is indeed quite wide-ranging as it spans biology and ecology, social theory, cultural anthropology and social geography. By tracing its humble beginnings, dating from around the 1860s, I would like to invite the reader to explore how the different disciplinary approaches, to varying extents, all contributed to preparing the ground for the pioneers of 'industrial metabolism' in the late 1960s. Since then it has gradually taken shape as a powerful interdisciplinary paradigm for empirically describing society–nature interaction.

Amusingly, this retrospective also shows us how willingly the social sciences absorbed the term metabolism from the natural sciences. Only later came the 'great divide' between these two disciplines, and the social sciences distanced themselves from naturalistic concepts. Only the awakening awareness of human-induced global environmental problems triggered again a change in social science thinking.

The term metabolism originated in biology and initially referred to the sum of all biochemical reactions a cell carries out each second. The purpose of metabolism then is to convert raw materials, obtained from the environment, into the building blocks of proteins and other compounds unique to organisms (Purves et al., 1992: 113). Eugene Odum (1973), a leading system ecologist, clearly proposes to use this term on every biological level from the cell to the ecosystem. In so doing, he challenged the notion of whether communities (the total of living organisms) or ecosystems have a degree of systemic integration comparable to individual organisms. These questions have since remained quite contested in biology. What is certain, however, is the fact that humans maintain a metabolism, drawing their energy from complex organic compounds (foodstuffs) that have been synthesised by plants from air and water, using the energy of the sun. Organic compounds are converted through oxygen from the air into carbon dioxide and water. If humans are to survive and reproduce, they must sustain their metabolism, if they cannot maintain this metabolic turnover, they will soon die out. From an ecosystem perspective, the concept of metabolism can be expanded to include material and energetic flows associated with living things (e.g., humans extract resources to build infrastructure).

Let us move on to the social sciences, where the concept of metabolism has equally found, at least in early sociological thought, its well deserved place. The social theoretical groundwork was laid by Marx and Engels in the mid nineteenth century as the founding fathers of modern social science were the first ones to apply the term metabolism (*Stoffwechsel*) to society. Influenced by the writings of their time (e.g., Moleschott 1852), they defined the labour process as a material exchange relation between man and nature. In so doing, their concept clearly emphasised an essential material interrelatedness between man and nature, thus challenging the widespread one-sided idea of man simply exploiting nature. Likewise, Herbert Spencer's ideas were strongly spurred by the advances in evolutionary theory and its implications for societal progress. He claimed that societal development was inherently linked to energy surplus. To his mind, energy alone would enable social growth while at the same time providing room for cultural deeds beyond basic needs. An interesting contribution also came from Ostwald (1912: 85), who argued that minimising the loss of crude energy and creating useful energy should be the base for societal progress. His judicious statement about the proper use of energy – don't waste energy, use it – could equally be taken from a contemporary IHDP[2] publication.

[2]IHDP stands for the International Human Dimension of Global Environmental Change Programme, an international, interdisciplinary science programme that is dedicated to promoting the human dimension of global environmental change (www.ihdp.unu.edu).

At the turn of the twentieth century, Sir Patrick Geddes (1885 in Fischer-Kowalski 1997) developed a type of three-staged input–output table based on the calculation of material and energy flows. To his mind, three transformation processes took place: extraction of fuel and raw materials, manufacture, transport and exchange. As he estimated the material and energy losses for each of the process stages, he laid the groundwork for later empirical analysis of societal metabolism. With sociology gaining ground as a growing discipline, material dimensions of society became largely silenced. It even seems that sociologists from Frankfurt to Chicago tried hard to remain in the abstract of the general study of society, remaining completely devoid of material features.

Similar to sociology, evolutionary theory was also strongly propagated in early cultural anthropology. Just like Herbert Spencer from the sociological strand, the prominent anthropologist Leslie White reawakened interest in energy as the propelling mechanism for socio-cultural progress. He stated that the evolution of culture was inherently linked to the capability of not only amassing energy yet also maximising its effectiveness through the use of technology (Singh 2003). A specific approach to the distribution of resources within the environment can be illustrated in the early study *Tappers and Trappers* (Murphy and Steward 1956), comparing two case studies of indigenous tribes, originally subsistence hunters and gatherers, who altered their ways of living due to a change in their metabolism. While not explicitly applying the term 'metabolism', the authors meticulously describe the metabolic transformation processes that have taken place in these societies. These were, on the one hand, the substitution of materials from their natural environment to materials based upon exchange with other societies. On the other hand, metabolic changes took place due to the complete replacement of certain materials and sources of energy by others. Both types of metabolic change inherently contributed to transforming many socio-cultural characteristics within their community fabric. Various other anthropological accounts on metabolism include authors like Harris (1966, 1977) and Rappaport (1971) who see the social organisation and culture of populations as functional adaptations of these communities to successfully exploit their environments. Latter is particularly worth mentioning, as he traces the energy flows of all activities involved in transforming secondary forest into a garden site. In so doing, he provided a detailed account of the energy exchange relation between the traditional slash-and-burn-farming Tsembaga Maring society and its domestic environment.

Finally, roots and traces of metabolism can also be found within the geographical and geological sphere. George Perkin Marsh, often referred to as the founding father of social geography, expressed his concerns about the irrational behaviour of men in destroying their base of subsistence, in his book *Man and Nature* published in 1864. Notwithstanding, it was not until Nathaniel Shaler's *Man and the Earth* (1905) came onto the global environmental stage, that the first concerns about the exhaustion of mineral and energy resources were voiced. He was particularly alarmed by the enormously rapid pace of global mineral resource consumption such as iron and copper. In Fischer-Kowalski's view (1997: 69), the change of focus from Marsh (1864) to Shaler (1905) somehow 'reflects the change in society's metabolism from an agrarian mode of production (...) to an industrial one'. In the midst of growing concerns about the limitation of our global mineral base, a 1955 conference convened a pool of 70 worldwide experts from different disciplinary arenas to discuss the growingly severe shortages of future energy and material supplies. All considerations, however, were still allotted to the input side of material metabolism. The 1960s marked quite a turning point within the metabolic discourse debate. When economic growth and environmental changes were becoming central issues within the scientific environmental debate, a new set of approaches was gradually developed to conceptualise and operationalise the metabolic concept. That was the analysis of material and energetic flows between societies and their natural environment. Contrary to neo-classical economics, which stress the importance of monetary flows and exchange relations, this approach embodies a biophysical conception of a society's economy and could hence bypass the shortcomings of money valuation (Giampietro and Mayumi 2000; Martínez-Alier 2002). Material and energy flow studies became ever more attractive since these newly emerging biophysical descriptions could be intricately related to the most powerful cultural system of modern society: the economy (see Fischer-Kowalski and Haberl 2007: 17).

Interested in the interdependence of human production systems, the American economist Walter Leontief developed the input–output method, which provided the base for a well-known tool in economics. Later, scientific attention on material and energy balances was drawn to investigating cities, as they offered a high concentration of population and commodities. In 1965, Abel Wolman was the first to attempt to conceptualise the metabolism of industrial societies as he undertook a case study of a hypothetical American city of one million inhabitants. Certainly a pioneering work on the use and consumption of goods, he linked the large amounts of wastes usually generated in a city to its inputs in order to establish per capita input and output flows (Wolman 1965). The study of urban metabolism has also intrigued other researchers who developed rather detailed processes of quantification of matter and energy throughput. In the 1970s, Newcombe and his research colleagues investigated the metabolism of Hong Kong, a mega-city which was experiencing a rapid transition period of high population growth and intense economic development. Around the same time, Brussels was also chosen for an assessment of all imports and exports of goods in and out of the city in order to establish an energy balance (Brunner and Rechberger 2002).

Departing from local case studies towards broader economies, Boulding (1966) illustrated the present world economy as an open system of energy, matter and information, which only provides for a certain amount of capital stocks. He therefore pleaded for a transformation from a 'cowboy economy', owing its success to maximising material throughput, into what he calls a 'spaceman economy' which would minimise the worldwide material throughput (Boulding 1966: 5). In 1969, Ayres and Kneese worked out a complete study of what was to be implemented as a material flow analysis of national economies in the 1990s. The novelty of their programme lay in the fact that they saw global environmental problems as being not primarily linked to economic growth,[3] but to the expansion of material throughputs of human societies (Ayres and Kneese 1969). These new insights, however, were only given

[3]As opposed to the Club of Rome publication *The Limits of Growth* by Meadows et al. (1972).

real attention in the 1990s, when common research interests began to focus on a holistic understanding of the physical aspects of economic processes, rather than merely looking at the output side. For the first time, the world realised that it was not only the toxicity of materials, but the total amount of material throughputs in an economy, that causes considerable environmental pressures. These novel insights instigated the development of new accounting methods and tools to attain a better understanding of the physical dimension of a socio-economic system. One such method is material and energy flow accounting which I am now to examine.

2.3.1.2 Material and Energy Flow Accounting

Material Flow Accounting (MFA)

The first methodological tool we are going to look into is material flow accounting (MFA) (Fig. 2.2), an accounting tool that traces all input/output material flows of a social system. A full material flow account commonly traces biomass, fossil fuel and mineral flows with only minor variations, depending on the specific characteristics of the system we are observing. Once the full material flow account is established, we are provided with different aggregated environmental pressure indicators for resource use. These pressures can either be detected at the society–nature interface

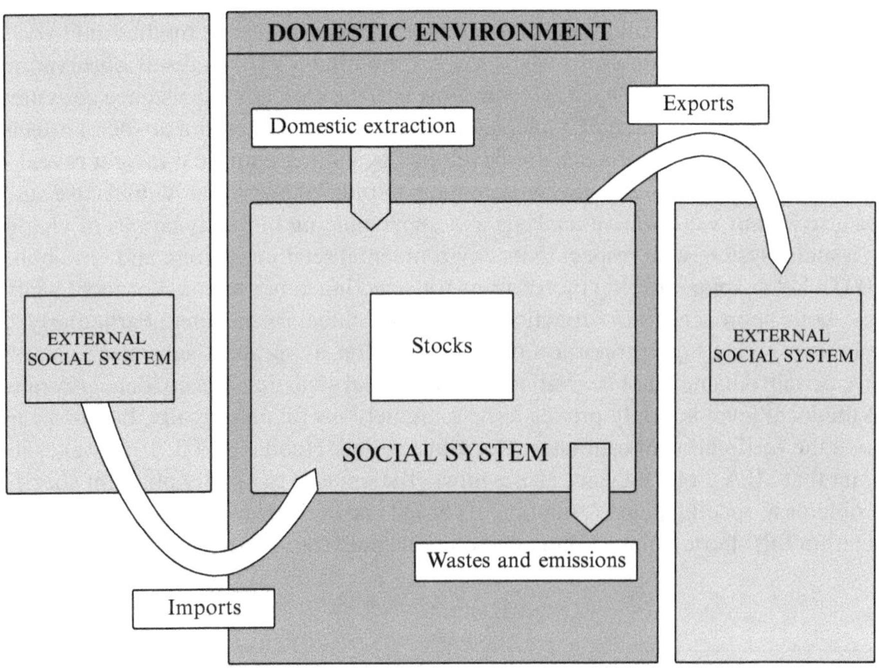

Fig. 2.2 System components and flow categories in MFA (adapted from Matthews et al. 2000)

(overexploitation of the natural system, for example) or between the social system under discussion and other societies (unfavourable import/export ratio, for example). National MFAs enjoy a much longer history than studies at the local level. National level MFAs are strongly promoted by Eurostat regarding it as one of the most powerful means for deriving environmental pressure indicators.[4] Macro-level studies are primarily concerned with two issues: on the one hand they attempt to analyse de-linking processes of material input from economic growth (Berkhout 1998; Hüttler et al. 1997; Schandl et al. 1999; Stern et al. 1996). On the other hand, they examine international material and energy exchanges between the global north and the global south (Giljum and Eisenmenger 2003; Muradian and Martinez-Alier 2001).

Contrary to the national level, local level MFAs have a much shorter history and less than a handful of empirical case studies actually exist so far: SangSaeng in Thailand (Grünbühel et al. 1999, 2003), Trinket on the Nicobar Islands (Singh et al. 2001; Singh 2003; Singh and Schandl 2003) and Nalang in Laos (Mayrhofer-Grünbühel 2004), with my empirical study being the most recent of all. With this in mind, it comes as no surprise to read about some of the stumbling blocks when engaging these tools in empirical local settings; tools that were originally designed for national accounting only. In this respect, we may consider local MFAs not as well elaborated methodologically, that implies dealing with all the difficulties that arise when adapting statistical tools to the realities of an empirical fieldwork situation. This endeavour undoubtedly requires the researcher's creativity as to what, when, how and to what extent to apply methods like counting, weighing, measuring, surveying, observing and sampling.

Yet national and local level MFAs are not necessarily driven by the same goals. National MFAs provide useful statistical information for societal self-observation; local MFAs, on the other hand, give insights into the society's subsistence activities, thereby providing a valuable understanding of specific material and social effects. How cultural coping strategies are developed is another empirical insight revealed by local MFAs that would otherwise remain untold. Likewise, individual case studies also permit a qualitative analysis and understanding of the dynamics of change to a social system with respect to its environmental relations (Singh and Grünbühel 2003). Yet despite differing motivations for selecting either national or local analyses, both approaches also function in a complementary manner. Particularly in countries with a high proportion of the population living from subsistence activities, certain data may not be available from official statistics. Hence, data generated at the local level not only proves to be extremely useful empirically, but also supports the verification of estimates. Notwithstanding, Hobbes (2005: 195) makes the point that MFA tools 'may not be the most efficient tool to rapidly pinpoint specific problems in specific places', blaming its broad and somewhat sedate systemic character for this shortcoming. Taking on board the therefore innate need for developing

[4] Austria was in fact among the pioneering countries to introduce regular national MFAs into the public statistical system.

additional indicators for local level research, she creates an interestingly detailed account of material flow categories and indicators in rural systems, encompassing the issues of agricultural transition and intensity, globalisation and food security.[5]

How To Do MFA Accounting?

Let us move on to examine the *modus operandi* of conducting an MFA. The first step to success lies in understanding the mutual relationship between environmental processes and socio-economic activities. Environmental pressure is exerted through a society's metabolism comprising 'the extraction of materials and energy, their transformation in the processes of production, consumption, and transport and their eventual release into the environment' (Schandl et al. 2002: 9). In other words, these processes mirror a society's economy. Prior to accounting the flows, the respective biophysical stocks 'belonging' to the social system need to be identified; and this requires the definition of two classes of system boundaries: the social–natural boundary, on the one hand, and the social–social boundary, on the other hand. From the perspective of the social focal system being studied, both boundaries are constantly reproduced by communication (by negotiation processes with other social systems about what forms part of the focal system culturally and physically) as well as labour (Fischer-Kowalski et al. forthcoming). Once the biophysical stocks – the human and livestock population, infrastructure and artefacts – are established, the boundary between the social system and its domestic environment becomes clearer. Accounting for the biophysical stocks is again a communicative process and defined also by the investment of labour. People, for example, may be politically part of a social system, but not engage in local economic activities due to their absence from the community (e.g., due to labour migration). Since they do not contribute to the metabolic profile of the social system under discussion, they cannot be considered part of the biophysical stock account. Generally though, accounting for the human and livestock population of a social system is still a much more straight-forward task than creating a stock balance for the, usually, high quantities of artefacts. Thus, in order not to get entangled with too many details, there is a general consensus among MFA experts to only include those artefacts in the stock balance that are actively used by the specific society for a period exceeding 1 year. Stocks are intricately linked to flows since they can only be reproduced by the flow of materials. The bigger the stocks, the more material flows are needed to sustain the stocks. Material flows are those materials extracted from the domestic environment or acquired from other social systems that are processed and consumed by the society and ultimately discarded back in the domestic environment or exported to other social systems.

Let us return to the two functional boundary distinctions and examine them in light of material flows. Material flows entering (inputs and imports) and leaving (outputs and exports) a social system go in accordance with these two system boundaries.

---

[5] Even though it would have been an interesting endeavour, Hobbes' proposed indicators were not used for the empirical analysis in Chapter 4.

As to the society/nature boundary, the first flow type concerns inputs (such as subsistence food) from the society's natural territory that a social system is entitled to exploit. This is often referred to as the domestic environment or resource base. As concerns the society/society boundary, the second type of material flows relates to imports (such as industrial foods or machines) from other social systems which enter the system through barter trade or purchase. Once inputs have entered the social system and been processed, transformed or simply consumed, they become outputs and are deposited onto the environment in the form of wastes and emissions. Similarly and in analogy to imports from other societies, there also exist exports to other societies such as traded goods for external markets. For a society that still has a high growth potential in material terms (i.e., a remote village community), the amount of material input into their system must be essentially higher than its material output onto the environment, since a portion of these input flows remains in the social system and is transformed into material stocks.

The types of material flows in MFA are generally reported in different groups constituting biomass, minerals, fossil fuels and manufactured products. Particularly for local empirical research ventures I suggest to add a fifth dimension, namely water. The use of water at household and source level may provide useful insights for describing transition processes. Biomass flows represent the most crucial elements to be examined, including food (for humans and livestock), the extraction of wood, straw, grass or any other useful materials taken from a society's domestic environment. The calculation of biomass flows equally entails accounting for the import/export of livestock as well as migration patterns of the local human population. Minerals usually comprise a large flow, as they often serve as building materials (e.g., cement, sand, clay and metals). Fossil fuels comprise another set of material flows. Since they are almost always industrially refined, they are accounted for as imports into the social system. By and large, the most significant are gasoline, kerosene, coal and petroleum and are mostly used in cooking, heating and for mechanical labour. Finally, manufactured products enter the system as imports, remain there as stocks for some time and leave the system again as wastes. In societies like the one examined in this book, such examples range from agricultural tools and machines to household appliances and transport vehicles.

Reaching a full material flow balance by integrating the input and the output sides stands at the core of MFA and can be achieved through applying the first law of thermodynamics: total inflows into a system equal total outflows corrected by changes in stocks. The importance here is that the law of thermodynamics equally relates to all separate entities (subsystems) within the social system, all of which are operating their own metabolism or input–output relationships. Apart from the input flow categories discussed above (biomass, minerals, fossil fuels and manufactured products), hidden flows may also be accounted for. The first kind of hidden flow derives from domestic material extraction and relates to materials intentionally mobilised, yet never used (e.g., by-products from mining or agricultural harvest). These products all cause additional environmental pressures for the domestic environment. The second hidden flow category exclusively occurs beyond the local level; it refers to trade and describes the environmental impacts that are linked to the international division of labour. Calculating these indicators in relation to imports and exports allows for the tracing of the shifting environmental impacts from one economic system to another. As for the empirical study of Campo Bello, no hidden flow categories were accounted for.

MFA Indicators

Once the individual material flow categories are calculated according to our two functional system boundaries, concrete environmental (un)sustainability indicators can be defined. These biophysical indices can be applied at all scale levels, from national to local and are broadly classified as input indicators, consumption indicators and output indicators. Whereas a whole range of literary sources (e.g., Adriaanse et al. 1997; Eurostat 2007; Matthews et al. 2000; Schandl et al. 2002) offer the whole array of MFA indicators, I simply present the ones I consider of great interest for the purpose of this study:

- *Weight of stock (WoS)* – the total weight of human-made artefacts in metric tons.
- *Direct Material Input (DMI)* – measures the sum of materials directly used in a socio-economic system, i.e., all materials that are utilised in production and consumption activities.
  DMI = Materials imported + domestically harvested
- *Domestic Material Consumption (DMC)* – measures the total amount of materials directly used in the socio-economic system. The term consumption includes both, intermediary resources needed for either domestic consumption of exports as well as those materials that are used for final consumption.
  DMC = DMI – exported materials
- *Domestic Processed Output* (DPO_{total}) – measures the total weight of materials, both from domestic extraction and imports, which have been utilised within the system's economy, before being discarded back to natural sinks. These flows occur at every level of the extraction, production/consumption and disposal chain. All flows are sub-grouped into emissions to air ($DPO_{to\ air}$), material loads deposited in water ($DPO_{to\ water}$) and materials dispersed into the environment ($DPO_{to\ land}$). Recycled material flows are not accounted for in this indicator.
  $DPO_{total} = DPO_{to\ air} + DPO_{to\ water} + DPO_{to\ land}$
- *Physical Trade Balance (PTB)* – measures the physical trade surplus or deficit of an economy and is equal to imports minus exports. For local studies, it is a useful indicator illustrating the extent to which the social system receives foreign resources or exploits its own resources for the benefit of other societies.
  PTB = Imported materials – exported materials

Energy Flow Accounting (EFA)

The second instrument within the methodological MEFA toolbox is energy flow accounting (EFA). EFA is an accounting method aiming to develop a complete energy balance of a social system: energy inputs, internal transformations, and outputs (Haberl 2002). Energy is accounted for in Joules (J) or its aggregate units (MJ, TJ) and used directly in the form of heat, light and power as well as human and animal nutrition. EFA is intricately linked to MFA inasmuch as it (1) uses the same functional system boundaries as MFA and (2) makes use of the same inflow, outflow and biophysical stock accounts already calculated for MFA. In so doing, MFA and EFA are complementary tools, indeed two readings of the same process.

EFA is based on the premise that non-material energy is usually detained in energy-containing materials and is used up by the social system when converted. Its comprehensive nature therefore allows for a complete energetic account stored in biomass, nutritional and work energy while equally revealing the amount of energy stored in those non-energetically used materials (such as construction wood or other building materials). It thus entails a complete series of conversion processes occurring in a social system. Energy initially enters a social system either through imports or domestic extraction. Both may occur for example in the form of biomass, fossil energy or nuclear fuel and are referred to as primary energy. Primary energy is subsequently converted into final energy, that is, for example, the processing of plant biomass into digestible human food. In an additional stage, final energy is translated into useful energy. In the case mentioned before this would imply the conversion of food calories into human or animal work. A comprehensive outline of these conversion processes sharpens our understanding of the total energy requirements of a society and equally provides a judgement for the effectiveness and efficiency of certain technologies used within the system (e.g., through calculating energy losses). At the end of the conversion chain, energy flows depart from the system either in the form of waste or are discarded prematurely in the form of heat dissipation or other material outflows. The application of EFA can be particularly useful for analysing societies in transition from one sociometabolic regime to another as it accounts for changes in energy flows, thereby revealing a certain metabolic profile of resource and land use patterns.

When applied to local studies, the following three accounting rules should be considered: firstly on the input side, energy inputs are calculated by counting all energy-rich inputs, immaterial energy flows (e.g., electricity, diesel) and biomass into the social system. The energy equivalent is assessed on the basis of its gross calorific value. Secondly, human and animal nutrition are both part of energy conversion processes and regarded as final energy inputs. Food inputs for animals raised to produce human food, however, are part of the conversion process from primary energy (e.g., plant biomass) to final energy (e.g., human food). Finally, human and animal labour both fall under the last stretch of the energy conversion process, namely that of useful energy.

EFA Indicators

In line with MFA indicators, a full energy balance derives at the following EFA indicators:

- *Domestic Energy Extraction (DE)* – measures the energy content of all biomass extracted domestically.
- *Direct Energy Input (DEI)* – measures the total amount of energy entering the socio-economic system, either by domestic extraction or import.
  DEI = Energy imported + energy domestically harvested
- *Domestic Energy Consumption (DEC)* – measures the total amount of energy directly used in the socio-economic system.
  DEC = DEI – exported energy

Yet, despite their great virtues and recognition in the international arena, providing no more than material and energy flow accounts would only reflect one side of the coin. Applied in a vacuum, they would simply fail to provide a holistic picture of the environmental pressures a society exerts, nor give any clues about the environmental quality of its resources. Here, the second set of metabolic relations comes into play: the concept of colonizing terrestrial ecosystems. It introduces a biophysical resource that is complementary to material and energy flows: the use of land and territory.

2.3.2 Colonization of Terrestrial Ecosystems

The concept 'colonization of terrestrial ecosystems' refers to the second compartment of our MEFA toolkit. It refers to a society's deliberate interventions into natural systems in order to render it more useful for them (Fischer-Kowalski and Haberl 1998, 2007; Krausmann et al. 2003). It has become operational in analysing land use and measuring the socio-economic pressures upon the environment. In many traditional societies, agriculture is the classical strategy to colonize nature. In the case of the Tsimane', for example, through their painstaking efforts and profound skills involved in making their labour effective, they have long successfully converted virgin forest territories into lush productive patches, executed and controlled by their own rules. Needless to say, for the colonized areas to remain of use value to the social system, it involves continuous hard work. The expenditure of labour, including human, animal and mechanical labour, has the transforming capacity of controlling environmental systems in such as way as considered useful for humans.

Colonizing strategies of natural systems are intrinsically linked to the exchange of energy and matter between humans and their environment as every mode of production embodies a specific set of metabolism and colonization strategies (Boyden 1992; Fischer-Kowalski and Haberl 1998; Sieferle 1997). Or to put it differently, the larger the population and the larger its metabolism, the more terrestrial ecosystems need to be colonized by the society in order to maintain this metabolism. Let us again consider agriculture and its dynamic nature. When a society changes from a traditional solar-based energy system (mainly based on biomass) to the use of new energy sources (e.g., fossil fuels), agricultural colonization activities equally become more complex and time demanding. With a change from biomass-based to industrial production modes, labour efforts to control the colonized systems intensify through the use of chemical fertilisers, herbicides and pesticides or the use of inanimate power.

Within the MEFA framework, the colonization concept analyses land use and land cover changes. Since socio-economic land use can be associated with changes in ecosystem processes, we can measure the actual impacts people have upon their territory. Or in other words, the loss of biodiversity caused by humans. This impact of land use can be measured by comparing ecosystem patterns and processes that would be expected without human intervention with those actually visible in the presence of interventions (Fischer-Kowalski and Haberl 2007). An example of this approach is the calculation of the 'human appropriation of net primary production' (HANPP) proposed by Vitousek et al. (1986).

2.3.2.1 Colonization Intensity Indicator: HANPP

HANPP, a term coming from systems ecology, assesses to which extent human land use changes the availability of energy in ecosystems. According to Haberl (1997: 47) HANPP can be defined 'as the aggregate effect of land use induced changes in productivity and biomass harvest on the energy availability in ecosystems'. The net primary production (NPP) is the amount of energy that primary producers – the plants – make available to the rest of living species. Measured in tons of carbon in energy units (MJ/m/year), it is estimated that today humankind 'co-opts' about 40% in terrestrial ecosystems (Martínez-Alier 2002). Changes in land use patterns induced by humans alter ecological energy flows and hence result in the loss of NPP. The higher the HANPP, the less biomass remains for 'undomesticated' biodiversity. With population growth, the proportion of NPP appropriated by humans is constantly increasing because of rising demands on land per person for growing foodstuff and timber.

Generating data on HANPP is accomplished in five stages and first of all requires the calculation of the NPP of the potential vegetation (NPP_0), i.e., NPP that would prevail in the absence of human intervention. In a second stage, the NPP of the actual or current vegetation (NPP_{act}) before harvest needs to be calculated. Thirdly, a calculation of the annual biomass harvest (NPP_h) needs to be carried out. In a fourth stage, calculating the remaining NPP (NPP_t) available in the ecosystem is done by subtracting the harvest from the current vegetation. In a final stage, HANPP can be calculated by subtracting the remaining NPP from the potential NPP.

In the process of calculating HANPP, we first have to clear the way of a few tumble stones. First, there is a technical question to be answered: should belowground primary production of vegetation units also be included? Measuring belowground productivity, though an important parameter, presents quite a difficult task since no reliable measurement instruments are actually available (Vogt et al. 1986; Waring and Schlesinger 1985).

Secondly, relations between energy flows, growth of biomass and biodiversity are also not so straightforward. Let us think of a desert area, for instance, which may have only limited biomass available, yet at the same time provides the habitat for numerous energy-rich endemic species. Due to these uncertainties and especially due to the lack of consistent data available, the HANPP analysis in Chapter 4 is merely based on the calculation of aboveground NPP. Assessing the NPP of potential vegetation of a region can be done via two differing approaches: the application of ecosystem models or bookkeeping models. The former entails the examination of climate variables, soil conditions and vegetation types among other variables; the NPP is modelled according to the use of these variables over time (Cramer and Fields 1999). With regard to the latter, studies of comparable vegetation units are evaluated (Ajtay et al. 1979; Lieth and Whittaker 1975). Likewise, the consultation of available maps showing the potential vegetation of the area under consideration would be an essential feature of the bookkeeping model. Actual land cover data or the actual NPP can be derived from statistical data and as well as remotely sensed data in the form of satellite imagery or aerial photography. Data on biomass harvest can either be obtained from agricultural and forestry statistics or, as is the case for the present study, from first-hand empirical data generated on the ground, on domestically extracted biomass.

2.3.3 *Functional Time Use*

So far, Sieferle (2001, 2003) has been instructive for the theoretical make-up of the sociometabolic transition framework as it stands now. He claims that in world history, socioecological systems (hunter and gatherers, agrarian regimes, industrial regimes) could be distinguished by the main energy sources social systems utilised for their survival – from passive to more active and intensive uses of energy. Each regime, in his view, could be characterised by similar land use practices, irrespective of the biogeographical conditions the society lived. While Sieferle's assumptions centre on energy constraints that push a system from one regime to another, different state through a process of transition, Boserup (1965, 1981) puts forward an interesting theory on the intensification of land use. In her examination of pre-industrial farming societies, she found that land use intensification, expressed in the frequency of cropping cycles, was triggered by human population growth. But Boserup has also been instrumental in linking the intensification of land use with labour time intensification. Together with other scholars (Harris 1977; Sahlins 1972; Wilkinson 1973) she holds the view that the work-load increases as we move from hunting-gathering forms of technology towards more land-use intensive forms of agriculture. The productivity of labour time, however, declines with land use intensification, since altering the nature of soils, moisture or topography (e.g., through irrigation or terracing) involves time demanding activities, both in terms of investment and maintenance (see Carlstein 1982: 351–353). The interrelation between population growth, increased work-load per worker and the intensified use of land use, which forms the core of her agricultural intensification theory which we are going to return to in Chapter 6, has been a useful point of departure for the incorporation of human time use into the theory of sociometabolic regimes.

2.3.3.1 Human Time Use and Social Metabolism

Human time has various metabolic characteristics. To start with, human time is a limited resource but evenly distributed among the human members of a social system. Especially in traditional social systems, the metabolic exchange relations between the people and their natural environment are coordinated by certain time norms (e.g., sexual division of labour) that are responsible for the functioning of the society. How human time is used, therefore, serves as a key to understand the social metabolism of a certain society. At the same time though, each human lifetime hour (from sleeping to wage work) can only be sustained through a certain metabolic input (matter or energy, for example). Failing to guarantee these inputs may lead to major tensions and specific solutions need to be found (see Fischer-Kowalski et al. forthcoming). These solutions may range from seasonal migration to demographic growth restriction measures and often do not go without periods of social conflict.

How can we establish the link between human time and a society's interaction with its natural environment? The answer is through labour time. This is a valid assumption for traditional societies, where every kind of interaction with the environment is exerted through the use of human labour. For hunters and gatherers, for example, the application of human labour is merely done in order to meet subsistence needs. If the amount of labour time available in the social system does not suffice to satisfy metabolic needs, or the energy return on labour too small, people will starve. At the same time, as only human labour is invested in the interaction with nature, the extent of human labour time is directly instrumental in impacts on the environment. Many studies show, however, that working time in hunting and gathering systems are generally low, showing a culturally stabilised 'leisure preference' (see, for example, Sahlins 1972). Hunters and gatherers have generally relatively low reproduction rates and invest comparatively little time in household chores. With regard to agrarian societies, working time is differentiated by season and class (upper classes are freed from subsistence work). As agrarian societies are generally characterised by higher reproduction rates, activities falling within the household and the economic system become much more time-consuming. Agrarian development is intricately linked to higher household chores as well as more work per unit area in order to obtain more output and enable the feeding of growing population numbers. Concerning industrial societies, Fischer-Kowalski (2007: 19) distinguishes between different phases. The first phase around the time of the Industrial Revolution brings with it an increase in working time. The second phase entails the creation of the 'leisure society' that is determined by low reproduction rates. Under the industrial regime, we no longer observe a direct link between the system's disposable labour time and its impact upon the environment.

In order to define the disposable labour time of a social system, I will apply the functional time use concept. For this purpose, human time is considered a key resource at the system level that is invested according to socio-cultural regulation mechanisms (e.g., labour division according to age and gender). To analyse these different socio-cultural regulatory functions and also take account of the time needed for physical reproduction, the concept entails a distinction between four different functional systems that each need time for their own reproduction. By and large, these categories can be divided into functions for personal and group reproduction, on the one hand, and socio-economic functions, on the other hand. The relevant subsystems we therefore distinguish are the person system, the household system, the economic system and the community system (for more detail, see Chapter 5). Such a holistic analysis provides a clear picture of the amount of labour time available at the system level, and, in so doing, aids our understanding of the specific opportunities and bottlenecks a society faces in its interaction with the natural environment (environmental burden). At the same time, analysing the time invested in each of the functional subsystems according to age and gender, gives an insight into the labour costs some of the age/sex groups may face (social burden).

Methodologically, the application of the functional time use concept simply lacks the experience we can draw upon with respect to MFA, EFA and HANPP. While Chapter 5 will deal with the methodological challenges that were encountered in the field situation when applying the concept for the first time, for now I would like to outline the two feasible approaches to data collection on human time use in the field: self-reported assessments, on the one hand, and direct observation, on the other hand. Concerning the former, applying self-reported assessments not only resolves most ethical issues, since the informant decides which behaviours to report and which not to mention, and moreover is rather time-saving for the researcher him- or herself. Notwithstanding, one of the major drawbacks relates to the problem of reliability. For example in a study of rural women's time use and after having collected both kinds of data simultaneously, Scheper-Hughes (1983) found that women failed to report 44% of their work as recorded by direct observation. The main reason for doing so is probably not deliberate deception but rather that people tend to remember their own behaviour selectively, in terms of cultural models of significant activities. On a similar note, Pastore et al. (1999: 341–343) noted that through self-reporting, the working time of traditional farmers may be underestimated due to two reasons. This, on the one hand relates to the tendency of particularly male household head informants to underestimate the involvement of children and elderly people in productive activities, as it does not get the same consideration as the time invested by adults. In addition, the distinction between working and non-working activities often becomes blurred, particularly when opportunity costs of labour are low. Keeping informant diaries may have a higher degree of reliability, depending entirely on how committed informants are. Notwithstanding, in traditional societies particularly, people often lack the writing skills necessary to use this method. It should also be borne in mind that people in other societies have a different worldview and may not use watches. Let me move on to direct observation. This research method can be done by either following the subject around all day (Lee 1979) or by using spot checks.

Both methods are subject to the fact that 'observers may be biased by their own expectations of what they are looking for or by expectations about the behaviour of women or men' (Bernard 2006: 430).

2.4 Conclusion

This chapter has provided the theoretical underpinnings of the 'umbrella' framework of sociometabolic regime transitions and its respective operational tools for empirical study. As concerns the former, we have learnt that a transition towards sustainability requires new ways of understanding the dynamics and impacts of current development approaches and their implication for global sustainability. Our theoretical understanding focuses on the society–nature interface and is based on the premise that societies in different world contexts face similar possibilities and constraints when interacting with nature. When certain biophysical parameters impose constraints beyond a certain threshold, the system turns into a qualitatively new state. This often dynamic and probably messy process of change is called transition. For our theory to become applicable to real world cases, this chapter has further presented the MEFA framework toolbox. Opening this toolkit, we were introduced to three methodological concepts and their respective operational tools: (1) social metabolism along with its functional tools MFA and EFA, (2) colonization of terrestrial ecosystems along with HANPP, its key indicator for measuring colonizing intensities, and (3) functional time use.

With regard to the concept of social metabolism, the metabolic idea which originated in biology has been adapted metaphorically to explain the interaction between human societies and natural systems. The metabolic model is based on the notion that a social system not merely reproduces itself socially and culturally, but also physically by exchanging matter and energy with its natural and socio-economic environment. The physical accounting tools MFA and EFA have been instrumental in developing full material and energy account balances; especially for national level studies they have provided a valuable tool for societal self-observation in terms of biophysical growth. The most important environmental impact indicators derived from such accounts are classified in input indicators (DMI, DEI), consumption indicators (DMC, DEC) and output indicators (DPO).

Notwithstanding, MFA and EFA alone can only tell one part of the story in terms of environmental pressures and environmental quality concerns. This is where the notion of colonizing comes into play, a concept that has become operational for describing land use. It refers to the deliberate intervention of humans in natural processes in order to alter certain ecological parameters and make them more useful for human purposes. One such colonization strategy is the application of agriculture. HANPP is the operational tool and the key indicator for measuring the intensity of colonization strategies of ecosystems and can be calculated in five stages: it first requires the calculation of the NPP of the potential vegetation that would prevail in the ecosystem without human intervention. In a second step, the NPP of the actual or current vegetation before harvest needs to be established. Following this calculation the annual biomass harvest figures need to be derived (usually taken from the MFA). Calculating the remaining NPP available in the terrestrial ecosystem is achieved by subtracting the harvest figure from the current vegetation figure. HANPP can finally be calculated by subtracting the remaining NPP from the potential NPP.

This chapter finally illustrated the concept of functional time use, a relatively new and less elaborated notion of embedding the use of human time within the theoretical transition framework. Equally rooted in a system's approach, the use of time is analysed at the social system level. The advantage of this communal perspective is the analysis of the particular possibilities and constraints a community is facing in its interaction with the natural environment (environmental burden) as well as for the people among each other (social burden). The functional interrelation between demographic reproduction rates and the amount of disposable labour time remains at the core of the metabolic time use analysis. As far as the link to the sociometabolic regime transition theory is concerned, for hunters and gatherers a direct relationship between a society's available working hours and environmental impacts can be discerned. With agrarian regimes, time use profiles change as household chores and activities falling within the economic system become more time-consuming. As these regimes are generally characterised by higher demographic reproduction rates (often due to the need for more agricultural labour), work-loads are further intensified in order to feed the growing population. Finally, the industrial regime entails two phases: the first is characterised by an increase in working time; the second is determined by low reproduction rates. With the industrial regime pattern, the direct link between the system's disposable labour time and its impact upon the environment becomes removed.

The next chapter introduces the empirical setting that takes us to the Bolivian Amazon, where we are going to explore the world of the Tsimane'. The Tsimane' are an indigenous community that largely sustains itself from farming, fishing and foraging. Like many other of today's rural smallholder societies, they are currently undergoing a transition – albeit in its very beginnings – from agrarian modes to a more industrial pace of life.

References

Adriaanse, A., Bringezu, S., Hammond, A., Moriguchi, Y., Rodenberg, E., Rogich, D., et al. (1997). *Resource flows: The aterial basis of industrial economies*. Washington, DC: World Resources Institute.

Ajtay, G. L., Ketner, P., & Duvigneaud, P. (1979). Terrestrial primary production and phytomass. In B. Bolin, E.T. Degens, S. Kempe, P. Ketner (Eds.), *The global carbon cycle* (pp. 129–182). Chichester, New York, Brisbane, Toronto: John Wiley and Sons.

Ayres, R. U., & Kneese, A. V. (1969). Production, consumption and externalities. *American Economic Review, 59*(3), 282–297.

Berkhout, F. (1998). Aggregate resource efficiency: A review of evidence. In P. Vellinga (Ed.), *Managing a material world: Perspectives in industrial ecology* (pp. 348–364). Dordrecht: Kluwer.

Bernard, H. R. (2006). *Research methods in anthropology: Qualitative and quantitative approaches* (4th ed.). USA: Altamira Press.

Boserup, E. (1965). *The conditions of agricultural growth. The economics of agrarian change under population pressure*. Chicago: Aldine/Earthscan.

Boserup, E. (1981). *Population and technology*. Oxford: Basil Blackwell.

Boulding, K. (1966). The economics of the coming spaceship Earth. In H. Jarrett (Ed.), *Environmental quality in a growing economy. Essays from the sixth RFF forum* (pp. 3–13). Baltmore: John Hopkins Press.

Boyden, S. (1992). *Biohistory, the interplay between human society and the biosphere*. Paris: UNESCO and Parthenon Publishing Group.

Brundtland Report (1987). *Our common future*.

Brunner, P. H., & Rechberger, H. (2002). *Practical handbook of material flow analysis*. London: Lewis Publishers.

Carlstein, T. (1982). *Time resources, society and ecology, Volume 1: Preindustrial societies*. London: George Allen & Unwin.

Clark, W. C., & Dickson, N. M. (2003). Sustainability science: The emerging research program. *In PNAS, 100*(14), 8059–8061.

Cramer, W., & Fields, C. B. (1999). Comparing global models of terrestrial net primary productivity (NPP): Introduction. *Global Change Biology, 5*(Suppl.1), iii–iv.

European Commission (1998). Fifth framework programme: Putting research at the service of the citizen.

Eurostat (2001). *Economy-wide material flow accounts and derived indicators. A methodological guide*. Luxembourg: European Statistical Office.

Eurostat (2007). *Economy-wide material flow accounting. A compilation guide*. Luxembourg: European Statistical Office.

Fischer-Kowalski, M. (1997). Society's metabolism: On childhood and adolescence of a rising conceptual star. In M. Redclift & G. Woodgate (Eds.), *The international handbook of environmental sociology* (pp. 119–137). Cheltenham, UK/Lyme, US: Edward Elgar.

Fischer-Kowalski, M. (2007). *Ageing, time use and the environment*. Presentation at the workshop: Research foresight for environment and sustainability – Megatrends and surprises, EEA. Copenhagen, Denmark.

Fischer-Kowalski, M., & Haberl, H. (1998). Sustainable development: Socio-economic metabolism and colonization of nature. *International Social Science Journal, 158*(4), 573–587.

Fischer-Kowalski, M., & Haberl, H. (2007). Conceptualizing, observing and comparing socioecological transitions. In M. Fischer-Kowalski, H. Haberl (Eds.), *Socioecological transitions and global change: Trajectories of social metabolism and land use* (pp. 13–62). Cheltenham, UK, Northampton, USA: Edward Elgar.

Fischer-Kowalski, M., & Weisz, H. (1999). Society as hybrid between material and symbolic realms. *Toward a theoretical framework of society–nature interrelation. Human Ecology, 8*, 215–251.

Fischer-Kowalski, M., Haberl, H., Hüttler, W., Payer, H., Schandl, H., Winiwarter, V., & Zangerl-Weisz, H. (Eds.). (1997). *Gesellschaftlicher Stoffwechsel und Kolonisierung von Natur: Ein Versuch Sozialer Ökologie*. Amsterdam: Gordon & Breach Fakultas.

Fischer-Kowalski, M., Singh, S. J., Ringhofer, L., Grünbühel, C. M., Lauk, C., & Remesch, A. (forthcoming). Sociometabolic regimes in indigenous communities, in scale interactions with an industrial world context, and the crucial role of working time. A comparison of case studies. Submitted to *Human Ecology*.

Geddes, P. (1885). An analysis of the principles of economics. *Proceedings of the Royal Society of Edinburgh*, read 17 March, 7 April, 16 June, and 7 July 1884. London: Reprinted by Williams and Northgate.

Giampietro, M., & Mayumi, K. (2000). Multiple-scale integrated assessment of societal metabolism: Integrating biophysical and economic representation across scales. *Population and Environment, 22*(2), 115–210.

Giddens, A. (2002). *Sociology*. Cambridge: Polity Press.

Giljum, S., & Eisenmenger, N. (2003). North-south trade and the distribution of environmental goods and burdens: A biophysical perspective. *Journal of Environment and Development, 3*(2), 354–368.

Grünbühel, C. M., Schandl, H., & Winiwarter, V. (1999). *Agrarische Produktion als Interaktion von Natur und Gesellschaft: Fallstudie SangSaeng*. Vienna: Social Ecology Working Paper 55.

Grünbühel, C. M., Haberl, H., Schandl, H., & Winiwarter, V. (2003). Socio-economic metabolism and colonization of natural processes in SangSaeng village: Material and energy flows, land use, and cultural change in northeast Thailand. *In Human Ecology, 31*(1), 53–87.

Haberl, H. (1997). Human appropriation of net primary production as an environmental indicator: Implications for sustainable development. *Ambio, 26*(3), 143–146.

Haberl, H. (2002). Economy-wide energy flow accounting. In H. Schandl, C.M. Grünbühel, H. Haberl & H. Weisz, (Eds.), *Handbook of physical accounting. Measuring bio-physical dimensions of socio-economic activities. MFA-EFA-HANPP* (pp. 28–44). Federal Ministry of Agriculture and Forestry, Environment and Water Management, Vienna.

Harris, M. (1966). The cultural ecology of India's sacred cattle. *Current Anthropology, 7*, 51–59.

Harris, M. (1977). *Cannibals and kings: The origins of cultures*. New York: Random House.

Hobbes, M. (2005). Material flow accounting of rural communities: Principles and outcomes in South East Asia. *International Journal of Global Environmental Issues, 5*(3/4), 194–224.

Hüttler, W., Payer, H., & Schandl, H. (1997). *MFA Austria: Methods, empirical results and reflections on delinking between economic growth and material turnover*. Wuppertal: Wuppertal Special.

Jenkins, R. (2002). *Foundations of sociology*. London: Palgrave MacMillan.

Juma, C. (2003). The global sustainability challenge: From agreement to action. *International Journal Global Environmental Issues, 2*(1/2), 17–32.

Kates, R. W., & Parris, T. M. (2003). Long-term trends and a sustainability transition, Retrieved April 10, 2009, from www.pnas.org/cgi/doi

Krausmann, F. Haberl, H., Schulz, N. B., Erb, K. H., Darge, E., & Gaube, V. (2003). Land-use change and socioeconomic metabolism in Austria. Part I: Driving forces of land-use change: 1950–1995. *Land Use Policy*, 20(1), 1-20.

Krausmann, F., Fischer-Kowalski, M., Schandl, H., & Eisenmenger, N. (2008). The global socio-ecological transition: Past and present metabolic profiles and their future trajectories. In *Journal of Industrial Ecology*, accepted for publication, 12(5/6), 637–656.

Lee, R. B. (1979). *The !Kung San: Men, women and work in a foraging society*. London: Cambridge University Press.

Lenski, G. (1974). *Human societies: An introduction to macrosociology*. New York: McGraw-Hill.

Lieth, H., & Whittaker, R. H. (1975). *Primary productivity of the biosphere*. Berlin, Heidelberg, New York: Springer.

Luhmann, N. (1984). *Soziale Systeme: Grundriß einer allgemeinen Theorie*. Hamburg: Suhrkamp Taschenbuch.

Luhmann, N. (1995). System and function. In T. Lenoir & H.U. Gumbrecht (Eds.), *Social systems.* (pp. 12–58). Stanford: Stanford University Press.

Marsh, G. P. (1864). *Man and nature; or, physical geography as modified by human action*. New York: Scribners and Sampson Low.

Martínez-Alier, J. (2002). *The Environmentalism of the poor: A study of ecological conflicts and valuation*. Cheltenham, UK, Northhampton, USA: Edward Elgar.

Matthews, E., Amman, C., Fischer-Kowalski, M., Bringezu, S., Hüttler, W., Kleijn, R., et al. (2000). *The weight of nations: Material outflows from industrial economies*. World Resources Institute: Washington, DC.

Mayrhofer-Grünbühel, C. M. (2004). *Resource use systems of rural smallholders. An analysis of two Lao communities*. PhD thesis, University of Vienna.

Meadows, D. L., Meadows, D. H., & Randers, J. (1972). *The limits to growth*. New York: Universe Books.

Moleschott, J. (1852). *Der Kreislauf des Lebens*. Mainz: Von Zabern.

Muradian, R., & Martinez-Alier, J. (2001). South-north materials flow: History and environmental repercussions in innovation. *The European Journal of Social Sciences, 14*(2), 171–187.

Murphy, R. F., & Steward, J. H. (1956). Tappers and trappers: Parallel process in acculturation. *Economic Development and Cultural Change, 4*(4), 335–355.

Odum, E. P. (1973). *Fundamentals of ecology* (3rd ed.). Philadelphia: Saunders.

Ostwald, W. (1912). *Der energetische Imperativ*. Leipzig: Akademische Verlagsgesellschaft m.b.H.

Parris, T. M., & Kates, R. W. (2003). Characterizing a sustainability transition: Goals, targets, trends and driving forces. Retrieved April 10, 2009, from www.pnas.org/cgi/doi

Pastore, G., Giampetro, M., & Ji, L. (1999). Conventional and land-time budget analysis of rural villages in Hubei province, China. *Critical Review in Plant Sciences, 18*(3), 331–357.

Purves, W., Orians, K., Gordon, H., & Heller, H. C. (1992). *Life: The science of biology*. Sunderland: MA: Sinauer.

Rappaport, R. A. (1971). The Flow of energy in an agricultural society. *Scientific American, 224*(3), 117–132.

Sahlins, M. (1972). *Stone age economics*. London: Tavistock.

Schandl, H., Hüttler, W., & Payer, H. (1999). Delinking of economic growth and materials turnover. *Innovation – The European Journal of Social Sciences* 12(1), 31–45.

Schandl, H., Grünbühel, C. M., Haberl, H., & Weisz, H. (2002). *Handbook of physical accounting. measuring bio-physical dimensions of socio-economic activities. MFA-EFA-HANPP.* Vienna: Federal Ministry of Agriculture and Forestry, Environment and Water Management.

Schellnhuber, H. J. (2003). Coping with earth system complexity and irregularity. In W. Steffen, J. Jaeger, D. Carson, & C. Bradshaw (Eds.), *Challenges of a changing earth* (pp.151–156). Berlin: Springer.

Schellnhuber, H. J., & Wenzel, V. (1998). *Earth system analysis. Integrating science for sustainability*. Berlin, Heidelberg: Springer.

Scheper-Hughes, N. (1983). Confronting problems of bias in feminist anthropology. In N. Scheper-Hughes (Ed.), *Women's studies* 10 (Special Issue). Philadelphia: Routledge, 126–149.

Shaler, N. S. (1905). *Man and the earth*. New York: Duffield and Co.

Sieferle, R. P. (1989). *Der unterirdische Wald: Energiekrise und Industrielle Revolution*. Hamburg: C.H. Beck Publishing House.

Sieferle, R. P. (1997). *Rückblick auf die Natur: Eine Geschichte des Menschen und seiner Umwelt*. Munich: Luchterhand.

Sieferle, R. P. (2001). *The subterranean forest. Energy systems and the Industrial Revolution*. Cambridge: The White Horse Press.

Sieferle, R. P. (2003). Sustainability in a world history perspective. In B. Benzing & B. Herrmann (Eds.), *Exploitation and overexploitation in societies past and present* (pp. 123–142). Münster: LIT.

Sieferle, R. P., Krausmann, F., Schandl, H., & Winiwarter, V. (2006). *Das Ende der Fläche: Zum Socialen Metabolismus der Industrialisierung*. Cologne: Böhlau.

Singh, S. J. (2003). *In the sea of influence: A world system perspective of the Nicobar islands*. Lund: Lund University.

Singh, S. J., & Grünbühel, C. M. (2003). Environmental relations and biophysical transition: The case of Trinket island. *Geografiska Annaler, 85 B*(4), 187–204 (distributed by Blackwell: UK & USA).

Singh, S. J., & Schandl, H. (2003). Socio-economic metabolism in the Nicobar islands: Empirical research in society–nature interactions. In B. Benzing & B. Herrmann (Eds.), *Exploitation and over-exploitation in societies past and present* (pp. 169–184). Münster: LIT.

Singh, S. J., Grünbühel, C. M., Schandl, H., & Schulz, N. B. (2001). Social metabolism and labour in a local context: Changing environmental relations on Trinket island. *Population and Environment, 23*(1), 71–104.

Stern, D. I., Common, M. S., & Barbier, E. B. (1996). Economic growth and environmental degradation: The environmental kuznets curve and sustainable development. *World Development, 24*(7), 1151–1160.

Turner, B. L., Matson, P. A., McCarthy, J. J., Corell, R. W., Christensen, L., Eckley, N., et al. (2003). A framework for vulnerability analysis in sustainability science. *Proceedings of the National Academy of Science, 100*(14), 8074–8079.

United States National Research Council, Board on Sustainable Development. (1999). *Our common journey: A transition toward sustainability.* Washington, DC: National Academy Press.

Vitousek, P. M., Ehrlich, P. R., Ehrlich, A. H., & Matson, P. A. (1986). Human appropriation of the products of photosynthesis. *BioScience, 36*(6), 368–373.

Vogt, K. A., Grier, C. C., & Vogt, D. J. (1986). Production, turnover, and nutrient dynamics of above- and belowground detritus of world forests. *Advances in Ecological Research, 15*, 303–377.

Waring, R. H., & Schlesinger, W. H. (1985). *Forest ecosystems, concepts and management.* Academic Press: New York.

Wilkinson, R. G. (1973). *Poverty and progress: An ecological model of economic development.* London: Methuen.

Wolman, A. (1965). The metabolism of cities. *Scientific American, 213*(3), 178–193.

Chapter 3
Exploring an Indigenous World in the Bolivian Amazon: The Case of the Tsimane'

Abstract This chapter provides the biogeographical, historical and socio-cultural background to the empirical case that follows in Chapter 4. I argue that this detailed account is necessary in order to provide a suitable context for understanding and interpreting the ensuing metabolic findings. For this endeavour, I am going to start with a general description of the indigenous Tsimane' territorial setting, and move on to a description of the natural and cultural landscape that shapes their world. The following section gives an overview of the most decisive historic events surrounding their ethnic community and entails an account of the peoples' growing exposure to outsiders. The final part of this chapter deals with the social sphere of the Tsimane' world. This involves a description of population trends as well as introduces the reader to life at the household as well as community level.

3.1 Introduction

Still today, the Tsimane' are an indigenous community that largely sustains itself from farming, foraging and trade. Despite contacts with outsiders throughout their course of history, it was not until fairly recently that their modes of subsistence have started to undergo changes. According to our theoretical framework, as societies develop, they change their social metabolism and the way they interact with and transform their natural environment. While the transition study framework is guided by the overall intent of identifying common systemic and metabolic characteristics between individual social systems, this book is equally concerned with discovering the individual opportunities and constraints of Campo Bello, the Tsimane' village community we are going to analyse in the forthcoming two chapters. This chapter to a certain extent provides the spadework for reaching both goals; it firstly provides the local environmental, historic, socio-cultural and political setting against which our empirical case study can be interpreted. On the other hand, this background information serves as useful when comparing the systemic features of different local systems. In such light, the overall aim of this chapter is to illuminate

L. Ringhofer, *Fishing, Foraging and Farming in the Bolivian Amazon: On a Local Society in Transition*, DOI 10.1007/978-90-481-3487-8_3,

the Tsimane' world from different angles; all of these are small but essential stones in the multifarious mosaic that constitutes their world. With these features slowly unfolding, I aim to gradually build a sound foundation upon which we can firmly base the sociometabolic analysis of the following chapters.

This chapter comprises four sections. It opens up with a general description of the Tsimane' territorial setting and its biogeographical features and then moves on to examine the parallel existence of the natural and cultural landscape surrounding the Tsimane'. Both domains are intricately merged and determine to a large degree the use of natural resources. A historic detour then guides the reader through a passage of growing exposure and contact with outsiders. It is a history that spans from early sporadic encounters to more permanent and increasingly difficult relationships with outsiders and points to the different social and ecological changes that were triggered along the way. In a final section, I foreground and examine the social sphere of the Tsimane' world, encompassing demographic trends and the peripatetic nature of settlement and community life. Such a rather detailed ethnographic sketch, so I believe, deepens our understanding of the system-specific challenges and opportunities of contemporary non-industrial communities which are currently all – albeit with different intensities – undergoing transition processes.

3.2 The Tsimane' Habitat

The Tsimane' inhabit a huge territory in the Beni lowlands covering 1 degree of latitude (from 14°5′ south to 15°5′ south) and 2° of longitude (between 66°5′ and 67°5′ west). It is some 50 km to the east of the first foothills of the Andean mountains to where it extends from gallery forest areas into the ample savannas of the Llanos de Moxos, providing a mosaic landscape of forests, open woodlands and savannah formations (Ellis and Araúz 1998). Two mountain chains dominate the area: Marimonos and Eva Eva. Their territory is bounded on the southwest by the Rio Quiquibey and the Rio Matos-Dumi, Rio Chevejecure and Rio Cuverene on the west. The Rio Maniqui that cuts its winding way along the outskirts of the municipal town of San Borja can be considered the very heart of the Tsimane' territory. It is along the meanders of this river and its numerous tributaries where most Tsimane' communities are found. There, they are somewhat sheltered from the outside world, yet close enough to the savannah area that extends from the town of San Borja. From the air, this riverine micro-region reveals a few inhabited clearings in the midst of the dense gallery forest, sometimes more, other times less visible to the human eye. Others but by far fewer communities are found along the Yucumo-Rurrenabaque road and the Rio Curibaba, Rio Matos, Rio Quiquibey and Rio Apere, all of which become the Rio Madeira after its confluence with the Rio Mamoré along the northeastern fringes of the Beni department. Crossing the border to Brazil, the Rio Madeira eventually becomes an important tributary to the Rio Amazon.

Today, the area inhabited by the Tsimane' is legally divided into 3 indigenous territories: the Indigenous Tsimane' Territory (TICH), the Multiethnic Territory (TIM) and the Indigenous Territory of Pilón Lajas (TILPA). The Tsimane' Indigenous Territory was titled in 1990 and has carried the official status of *Tierra Comunitaria de Orígen* (TCO) since 1997. The latter is the product of a long and complex process and recognises the official use of the natural resources by the Tsimane' within the political boundaries of their territorial land. The TICH covers a surface of 401,322,805 ha, an area half the size of Corsica. The vast majority, in fact almost 90% of the Tsimane' population lives within the boundaries of the TICH, while the remaining 10% dwell in the TIM where they share their land rights with other indigenous groups like the Movimas, Yuracaré and Mojeños[1] (Tobías 1998).

3.2.1 Biogeographical Features

This part of the Bolivian Amazon basin that is home to the Tsimane', harbours a huge variety of soils and soil compositions. More than their mostly non-indigenous neighbours, the Tsimane' are mindful of the topographical and pedological diversity of their territory. General empirical understanding of their surroundings is rooted in centuries of observation and agronomical experimentation which has provided them with a detailed knowledge base of the various components of their environment. Let us look at soil, for example. The Tsimane' typically associate a specific type of soil with its specific texture, while soil colour and to a lesser extent, drainage tend to be used as secondary modifiers for soil classification. In their taxonomic system, soils are classified into 7 primary categories that sometimes take on regional variations (Piland 1991). The clay-dominated *jak pirij* (sticky earth) and *jak bojca* (slippery earth) constitute the first grouping; *jak tsincus* (black earth) has a thin and dark humus surface horizon and is quite popular for its fertility; *jak tsijtyi* soil types with their high alkalinity content are often used as salt licks by game animals; *jak chuñus* (red earth) is usually found in the upstream Rio Maniqui on high river banks, while *jak jaimay* (beach earth) and *jak puijdaij* (powder-like earth), both dominated by sand, are more prevalent along lower beach areas. Tsimane' men, when asked to identify a particular soil, skilfully take a sample from the ground and slowly grind it between their thumb and index finger, often picking up an additional sample to affirm their presumption. As if naturally ingrained in their profound environmental knowledge base, Tsimane' boys would commonly shout the name of the soil by just standing on it. Tsimane' women, on the other

[1]The available data from 1998 (Tobías 1998) reveals that 75 Tsimane' communities were distributed along the Rio Maniqui (35 along the Lower Maniqui, 40 along the central and upper parts of the river), while merely 11 Tsimane' communities were found to live within the TIM.

hand, tend to use a more utilitarian system for describing a particular soil; they simply name the soil for the crop they feel will be most suitable to plant on that particular stretch of land.

When reading ecological literature about the Amazon, one cannot help but be discouraged by the mounting evidence of poor soils found by a number of researchers, all somewhat lamenting the soils' general acidity, low fertility and varying degrees of aluminium toxicity (see, for example, Sanchez 1976). At least my lush image (and I suppose that I am not alone here) of the Amazon forest had been that of a giant green expanse brimming with fecundity, fertility and wilderness. It was hence not surprising, when the results of a large-scale FAO soil evaluation in the Tsimane' territory south of San Borja carried out in 1979 confirmed fairly low contents of lime, potassium and phosphorus and the absence of carbonates in the regional forest soils. Concerning acidity, pH-values range from 4.5 in the superior horizon to 5.6 in the deeper soil layers. From this we can deduce that soils in the region sustain a verdant forest only through a rapid uptake of nutrients from decaying organic matter in a thin topsoil layer; clearing the forest for horticulture exposes the soils to degradation (Richards 1985; Roosevelt 1989). In contrast to relatively light forest soils, savannah soils are extremely loamy and heavy in texture. Their compact nature impedes proper drainage and makes forest vegetation impossible, a fact that frequently leads to recurring seasonal flooding (Hanagarth and Sarmiento 1990).

Similar to other Amazonian regions, the regional climate is generally warm and humid. Still, there are marked seasonal climatic differences, registering temperature ranges from a minimum of 19–22°C to a maximum of 30–31°C. During my first research spell in 2004, the data I had gathered from the meteorological station located at the local airport in San Borja registered the highest temperature as 28.1°C in November and the lowest temperature as 20.8°C in July. The most extreme climatic values ever recorded at the station were 39°C in October 1987 and 6°C in July 1985. Atmospheric humidity varies little and average annual rainfall ranges between 1,870 and 2,550 mm (CIDDEBENI 2002). The general pattern is for the day to warm rapidly until late morning, reaching temperature peaks in mid-afternoon. Frequently accompanied by a steady and cooling breeze, it is a time of communal rest and people seek the generous shade of a lush fruit tree.

As in most parts of the Beni region, there are 2 distinct annual seasons in the Tsimane' territory: the dry and relatively cold winter between May and November, and the wet and humid summer between December and April. During the January peak rains are frequent, often accompanied by storm fronts. High morning temperatures favour air convection and massive cumulo-nimbus clouds gradually build up towards the late afternoon. Often, the storms do not break immediately and during this period thunder might rumble for days without producing even a drop of rain. When storms eventually do break, one can experience a sudden downpour of several inches of rain within just a few minutes. By then, the Tsimane' have hopefully sought refuge in their homes, as the forceful winds accompanying the heavy rains have the power to tumble top story trees with a medium height of between 40 and 50 m. Except for extremes, clouds and rain usually pass swiftly and give way to blue skies and drifting white clouds. The rains practically invite molesting swarms of mosquitoes onto the scene, leaving everyone covered in itchy red spots. A characteristic scene with the Tsimane' is for people to abruptly interrupt whatever they are doing to slap at an exposed part of their body and later examine their palm for traces of mosquitoes. It is frankly a true relief when they (nearly) disappear with the onset of the dry season in May.

The dry winter season is marked by occasional cold spells or *surazos* that often accompany the strong winds. This is a typical feature of the whole Amazon region in the winter, as frigid air masses sweep north from Antarctica (Johnson 2003: 16). Winds from the north predominate all year round, while winds from the south and southeast are common from April through to August. During these spells, winds sometimes even exceed 150 km/h (CIDDEBENI 2002). These cold spells are repeatedly accompanied by a light rainfall locally called *chilchi*. With the onset of these winds, the sun becomes obscured by heavy clouds and stiff breezes blow through the well-ventilated Tsimane' houses, some of which are still traditionally constructed with no or just one wall made of uva grass (*Gynerium sagittatum*). These are generally hard times for the people as many Tsimane' do not own warm clothes to shield them from the cold. Getting up numerous times during the night to add more logs to the embers of the slow-burning fire seems the only means to keep warm. Although the strong winds tend to become increasingly moderate after a short while, the cold spells may last up to a week, before they give way to a warmer climate again by mid-August.

3.3 Landscape and Cosmos: Bridging the Natural and the Cultural

According to the theoretical framework guiding this monograph, social systems are conceived as hybrids, a structural coupling of a cultural system with certain sets of material components. Both overlap and constitute the biophysical structures of society (see Fischer-Kowalski and Haberl 2007). The way a society organises these biophysical elements is intricately linked to its culture. I would argue that this is even more evident in more traditional societies, whose ways of interaction with nature are mostly dictated by strongly embedded traditions based on their own cosmology and cosmogony, passed down from older generations through oral tradition. Myths, rituals and taboos help and direct the interaction of people with their natural environment as they provide the legitimacy for relationships and regulatory rules for resource use. While delving into the wide-ranging Tsimane' spiritual world would go beyond the scope of this book, I have decided to depict those features of the Tsimane' cosmic worldview that combine the notion of biophysical landscape and cosmos. This means a brief encounter with the Tsimane' cosmic landmarks before examining the peoples' use of the forest, the river, and the farming site.

The works of cultural anthropologists like Riester (1976), Daillant (1994), Ellis (1996) and most recently Huanca (2006) are probably the most intriguing sources which introduce a Tsimane' world that is charged with supernatural meanings. According to Tsimane' narrative, people, animals, plants and stars were originally members of the same society, sharing the same language and cultural traits. Still today, a deeply ingrained cultural world coexists alongside the natural world of the Tsimane' and numerous myths and widely known oral histories are an active part of daily community life. The Tsimane' regard for their natural environment rests upon two concepts. First, there is a strong belief that natural and spiritual beings guide the relations between humans and nature, thereby denying the existence of any impersonal forces. Secondly, inappropriate human behaviour and disrespect for forest resources provokes the anger and mistrust of the guardian spirits. If these rules are broken, sorcery powers are released that cause harm, illness and even death to the people and their families. In such a light, we may say that daily Tsimane' routine not merely involves material satisfaction, but also the constant awe and awareness to comply with all sets of rules guiding the relations between the natural and the supernatural world.

3.3.1 Cosmic Landmarks

The Tsimane' believe that their habitat contains 3 spaces which are a combination of geographical and cultural features: the *jacche'* (earth space), the *che've* (celestial space) and the *jaccan* (underground space). All of these 3 worlds are charged with mystic beings. These microcosms, of which some are considered dangerous while others refer to historic events that occurred there, have had an important regulatory role in forest, land and resource use. Trees, rocks and large rock outcrops are thus encountered with awe if not fear as they are usually dwelling places for spirits. Whereas today most Tsimane' travellers still recognise these potentially powerful places on earth, in the past they would try to avoid these places or, if passing them, try not to perform any rituals there. Two landmarks deserve attention. The first is *Pa'tsene*, a historic salt lick that is still today regarded as a powerful site. The second is the myth-woven Milky Way, a periodic marker that yet again bridges the cultural with the natural domain.

Pa'tsene is a salt spring that lies in the very centre of the Tsimane' earth space (Huanca 2006: 140). The cultural anthropologist Isabel Daillant (1994) remarks that until recently, drawings of vulvas were carved in the rocks nearby. These, however, were all destroyed by a logging company that entered the territory in 1996. Up until the 1960s, the majority of salt consumed by the Tsimane' would be obtained from this salt lick near the source of the Pachene river, some 60 km away from the market town of San Borja. In her work on the socio-cultural ethics of the Tsimane', Ellis (1996: 127) calls the site a 'spectacular assemble of large, moss covered rocky outcrops creating a tunnel-like effect' that is 'the cradle of the Tsimane' peoples'. Still today, these formations embody and emit strong powers coming from potent guardians living within. Even the water itself is believed to have healing powers. Today, however, it is mainly used as hunting ground since many animals are attracted by the rocky moss covered formations to lick the salty water.

In the past, entire Tsimane' families would make seasonal expeditions to *Pa'tsene* to collect salt. Yet a trip to the sacred place, normally taking place in August during the dry season when the salt lick was drier, implied an arduous journey. People canoed for 3–6 days upriver following the innumerable meanders of the Rio Maniqui. As the site is tucked away behind a hilltop, people had to walk for several hours, first along the riverbanks, and then climb the hill. Physical landmarks were indeed significant as primary forest patches, mountain tops, barely visible hunting trails or temporary camp sites guided the travellers throughout their passage. Along the way, people used to stop and catch fish or simply rest from the fatigue of the expedition. Upon arrival at the spring, knowledgeable elders would clean the salt lick before starting to process the salt. People would fill aluminium pots with water, place them over a fireplace and leave the water to boil until it all evaporated and the salt remained. The units of hardened salt processed were then wrapped in large leaves and taken home. Upon a man's return from the salt lick, his wife would serve him copious amounts of manioc beer to recompense for the strenuous voyage he had undertaken. With the rise of commercial salt, obtainable from the market town of San Borja since the 1960s, the rituals and taboos associated with the extraction of salt at *Pa'tsene* gradually diminished. While no longer really visited for salt extraction, the salt lick remains highly valued for its historic importance and is still today believed to be a dwelling place for guardian spirits.

The Milky Way is one feature representing the celestial space. It is yet another cosmic landmark and bears a strong meaning as a periodic marker in Tsimane' oral history. In the indigenous worldview, in ancient times the sky was close to the earth, with sky stones falling bit by bit onto the earth. When the sky darkened, it was believed to fall even further. In those times, there were no stars twinkling at night. According to Tsimane' myth, *noco*, a small lizard-like being, lived with an old woman and provided her with fish. As he grew older, he started to provide the fish for all the people. One day, *noco* left the earth to put the sky higher up and to sustain its altitude by himself. As he stretched his body resembled an arch, the Tsimane' suppose that this celestial arch has become the Milky Way. The sky has since never fallen again. Today, the Tsimane' still believe that *noco* inhabits the sky and when the Rio Maniqui dries out during the year, it is due to the benevolence of *noco* to ease the catching of fish. This can partly be explained in astronomical terms, since during the wet season the Milky Way runs parallel to the Rio Maniqui (north to south). An increase of river levels that might even lead to flooding hence coincides with the period when the Milky Way changes its position. Then during the dry season, the Milky Way traverses the Rio Maniqui, which is perceived by the Tsimane' as 'him' obstructing the river, causing it to dry up. Similarly, as the Milky Way can only be seen on very clear nights, it is during the rainy season, when the atmosphere is drenched in humidity, it becomes more complicated to spot (see Huanca 2006).

3.3.2 The Forest and Its Use

It is the combination of abundant solar radiation and high humidity that is particularly favourable to the growth of outstanding forest biodiversity. Except for occasional swamps, lagoons and savannah areas dotting the map of the wider region, thick und humid forest grows in an unbroken mantle over most parts of the area. The forest areas contain over 1,000 tree species, above all mahagony (*Swietenia macrophylla*), cedar (*Cedrela odorata*), oak (*Amburana cearensis*) and others such as chestnut or almond trees (*Berholletia excelsa*), palm hearts (*Euterpe precatoria*) and rubber (*Heavea brasiliensis*). What is typical of the region is the presence of three to four main stories of trees. The top story is formed of trees with average heights of between 40 and 50 m with straight trunks and spreading crown canopies. These gigantic tree species are particularly susceptible to sudden gusts of wind which wipe over the territory at certain periods over the year. The colossal and pyramidal trunks form wooden panels, from which Tsimane' men skilfully make mortars for squashing plantains or manioc. The middle story is the thickest with tree sizes of between 20 and 30 m. Palm species, such as peach palm, are a common feature of this level. At every level of the forest, entwining liana and ferns intermingled with young saplings of taller tree species form a dense canopy, whose fallen leaves provide the ground with fertile organic matter. The lowest story is made up of natural windfall clearings, fallow garden sites or all together abandoned horticultural plots. This story also provides a fertile hunting ground. Game resources are plentiful and forest dwellers include howler monkeys, spider monkeys, capybaras, squirrels, giant anteaters, long-nosed armadillos, tapirs, collared peccaries and red brocket deer. Other forest dwellers are toucans, parrots and various other birds and insects which all contribute to an enthralling wall of sound when strolling through the forest at night, especially at the edge of the forest near the meanders of the Rio Maniqui. From an ecological standpoint, the system ecologist Eugene Odum (1973) contends that this type of tropical rainforest is in a state of dynamic equilibrium, as its energetic exchange system is self referential, meaning that it operates as a closed system. Minerals and organic matter are constantly recycled by micro-organisms and bacteria, relying on the protective strata of forest plant cover.

Taking an exploratory wander into the unplumbed depth of the forest always meant a pure adventure when accompanied by one of my Tsimane' hosts. I was especially intrigued by the young Tsimane' boys and girls who would not only make me aware of the particular species in front of us, but on many occasions happily revealed the taboos and myths related to that particular type. This leads us to explore the cultural sphere of the forest and how it is used by the Tsimane'. While the forest dispenses its incredible resource wealth liberally, it obliges the Tsimane' to play by rules they do not always control. In order to tap into these resources, various distinct cultural rules are to be followed. Every plant and animal has its own spirit or guardian, an *a'mo*. For the Tsimane', the forest is generally regarded as a pleasant place, calling it 'new land, razed, cooked and clean' (Huanca 2006: 140), yet also full of dangers. It is not simply a nice place for a walk, one sets out into the forest for a definite reason and with a definite goal in mind.

The treatment of trees and plants is mediated by plant masters. Tomás Huanca's (2006) excellent analysis of oral testimonies allows for a distinction between tree species with fierce spirits and those somewhat less dangerous for the human forest visitor. As concerns the first group, the guardian spirit of these trees has the power to transform itself into a jaguar. Once transformed, these jaguars act just like their real counterparts from the forest. They roam about the forest freely, appear anywhere and suddenly attack humans. The spirit of the *vojshinaj* (Kapok tree, *Ceiba pentandra*) is particularly known for its strong powers and recently-born babies, menstruating women or young girls are particularly vulnerable to fall victim of its sorcery. Simply passing near this kind of tree might be a life-threatening endeavour. Yet the Tsimane' have acquired ways to protect themselves from these powers through the use of fire and smoke. It was quite a common scene to run into a Tsimane' family walking on a forest trail, usually with the man going ahead carrying one or two smouldering logs and the mother behind, gently covering her baby's face with her hands in order to protect it from the malign spirits inhabiting the surrounding area. Another group of trees does not have the transforming powers attributed to the first group. While they are also inhabited by spirits they do not transform themselves into jaguars. With some of these trees the Tsimane' share kinship and are especially careful to protect them from dying. On an interesting footnote, despite the fact that the Tsimane' view the forest as primarily a man's domain, they regard large trees as dwelling places inhabited by both male and female spirits.

Interaction with the animal world takes place through the communication with animal masters (*jäbäbä*). Animals are given human attributes and, so Tsimane' cosmology narrates, in former times animals were members of a society that also included humans. Master spirits are the owners of wild animals and humans must ask the master for permission before killing the animal. Hunting, as a central activity within the Tsimane' array of subsistence activities, is a fine example of demonstrating the requirement of two sets of skills: the technical knowledge and the understanding with the game and its guardian spirits that is based on the principle of magical complicity. To put it differently, it is the careful combination of practical hunting knowledge and the compliance with deeply ingrained ritual beliefs that make for a good and respected hunter among the Tsimane'. The technical skills and procedures involved in hunting entail detailed knowledge on the use of hunting weaponry, the mastering of hunting skills like tracking and stalking, thorough knowledge on the seasonal and spatial living habits of the prey and compliance with rituals to appease the spirits at hand.

An exclusively male activity, the fabrication of traditional hunting weaponry (hand-made bows and arrows) is becoming increasingly replaced by rifles and shotguns. Having said this, however, manufacturing bows and arrows is still practiced widely and a few men are particularly reputed for their skill in the area. In Campo Bello, Santos had gained a reputation as an expert craftsman and other men would naturally approach him for the production of quality arrows. Shotguns, evidently more efficient and time saving hunting devices, require ammunition which imposes a problem when one lives far from a commercial centre. Some remote communities, tucked away in the depth of the Tsimane' forest, only receive occasional visits from river traders, leaving the hunters entirely dependent on these unreliable and unequal circuits of exchange. Besides these active weapons, traps are sporadically employed as passive weapons. They are particularly useful devices for eliminating rodents on fallow sites or producing gardens. Dogs are also loyal hunting companions and are trained by their owners to kill certain animals that they hunt. The best times of the day to go on hunting trips are usually early mornings or the late afternoon, when both the day-time and night-time animals are either just waking up or settling down. Pregnant women have a particularly pronounced craving for juicy meat and their husbands have the imperative duty to fulfil these desires. A serious hunter thus leaves his home at the crack of dawn and stays out for most of the day. Some hunting trails allow for hunting trips lasting for several days, usually one day to stride out to the hunting site, one or 2 days in the hunting area and one day to return. Game is usually hunted within a 3 km radius from the Tsimane' village settlements (Apaza et al. 2002). Stalking and tracking are supposedly one of the most difficult but pleasurable hunting endeavours. The hunting of large ground game requires the skilful Tsimane' hunter to track characteristic prints and swiftly interpret the slightest clues. Expect for jaguars, sloths, bats and snakes most animals are edible for the Tsimane'. Snakes are considered undesirable since they are poisonous, too small to nibble on and, more importantly, have a rotten smell. The best hunting season occurs around April and May when animals are at their fattest. Alternatively, long periods of heavy rain extending from January to February have a nefarious impact on hunting, for the men usually remain at home and use the time to manufacture hunting devices or other commodities.

Besides these practical skills indispensable to assure a good hunt, compliance with hunting rituals and taboos have an equally strong bearing on the success of a hunt. The loss of hunting skills is sure evidence that some taboo has been broken. While some researchers have been intrigued by the intricate relationship between dreams and hunting success (e.g., Riester 1976; Chicchón 1992; Mayer Roca et al. 2000), most taboos are related to refraining from earthly pleasures. The night before a hunting trip, for example, a successful hunter must abstain completely from sexual relations. Strict dietary prohibitions depend on the animal to be hunted; eating hot peppers for example is forbidden before hunting tapirs. On several hunting occasions I observed that men, upon finding wild honey, would not consume it themselves but offer this succulent nectar to me. Yet I was unable to find out whether the consumption of honey was yet another dietary prohibition or simply one of their many nice gestures of hospitality. Among the Achuar, honey is supposed to clog the lungs of a hunter, thereby making it impossible to use the commonly used blowgun (Descola 1996). Upon the successful hunters' return, the meat is preserved through salting and drying the chunky pieces into *charqui* and manioc beer is generously shared among family and neighbours. When the meat is smoke-cured on a wooden platform built over the open fire, it will keep for about 10 days before it becomes infested with maggots and foul-tasting. The consumption of fermented manioc beer and the distribution of game meat to close family constitute a moment of celebrating the hunter's success and a token of gratitude for the animal master.

As opposed to hunting, gathering is a far less risky undertaking and indeed a fairly easy-going and entirely prosaic activity. Always a good excuse to break the monotony of the daily round, people are not reproached if the outcome at the end of the day is modest. Gathering seasonally available wild fruits like *ibijqui (Rheedia achachairu)*, sweet limes or other kinds of citrus fruits simply offers a welcomed variety to the Tsimane' diet, topping up the daily fare rather than replacing it. Forest fruits are rarely served with a meal but are considered special treats that are enjoyed throughout the day. While gathering takes place all year round as an accompanying activity to hunting or fishing, the season when most trees come into fruit extends from December to March. This is a period when women and children set out for little walks in the direction of known fruit trees or palms. Fruits are either beaten out of the tree with a long stick or simply picked up off the ground. Sometimes brave young boys climb up to shake the twigs and bigger branches and grasp what they can reach. Gathering is not only limited to edible plants, as the forest offers an enormous array of plant species for preparing medication, fish poisons, firewood and materials for housing construction.

Starting in August, Tsimane' women gather *cajñere*, fibres from the balsa seed pod. This has in recent years become quite a lucrative business as a local cooperative in San Borja has started to use these fibres, similar in texture to that of cotton, for stuffing pillows and mattresses. By the end of September, it seems that Tsimane' women of all ages in the area spend most of their daytime collecting and transporting the fluff to the cooperative. The foraging of animal products equally falls within the scope of this joyful activity. Wild honey collection, for example, is a common but fairly brave undertaking and honey collectors often return home covered in bee stings. Extracting honey also means quite a bit of work as thick trees are cut down with axes. As soon as the tree has fallen onto the ground it is the women's task to bring jars and fill them up with honeycombs. The turtle egg season spans from late August to early October every year, guaranteeing an abundant supply of animal protein during this period. They are quite easy to locate along the river beaches where turtles leave their eggs in nests. Once harvested, they can be dried in the sun in order to be consumed at a later stage. Reportedly, when preserved by drying they last for about 8 weeks without spoilage (Chicchón 1992).

3.3.3 The River and Its Use

For the Tsimane' the Rio Maniqui occupies an extremely important place in their daily lives and is considered a resource that serves multiple functions. To start with, not only does the river quench the thirst of the exhausted hunter on his way home from a hunting trip, but also provides the only water source for cooking in the absence of individual wells. Equally important, as a mode of transport it is navigable all year round. In this respect, as the river links remote Tsimane' communities to the commercial centre of San Borja, it eases their integration into the regional market economy. At certain times of the year, river traders embark on journeys up and down the Rio Maniqui with the intention of selling or bartering highly valued goods among the Tsimane' in exchange for rice or plantains. Adding to these functions, the river also has a social purpose. During the daytime it is a place where women meet to wash their clothes and children indulge in bathing at dawn, while at night, the setting may also serve for couples to engage in erotic play. Leaving the worldlier sphere, the river is also home to benevolent and malevolent guardian spirits that are all charged with supernatural powers.

Just like forest resources have their spiritual owners, cultural practices on the river setting are performed on a similar note. In the Tsimane' cosmic world, there are two significant forces that are in control of fish resources: *i'dojore* as the benevolent protector who assures that fish are always plentiful and *o'pito*, the unanimously feared fish guardian (see Pérez Diez 1983). Especially in former times, shamans had an important mediating role between the Tsimane' and *i'dojore* as they knew how to pay the fish owner in order to 'open the fish door and allow fish to go upriver' (Huanca 2006: 155). Older people narrate that the use of dynamite to catch fish, a popular method applied by colonists, would anger the guardian inasmuch as he would shut the fish gates and make the Tsimane' suffer. *O'pito* is the evil force that controls fish resources and the Tsimane' are in constant fear of his fierce powers and unpredictable spirit to cause harm and even death to others. In former times, if *o'pito* provoked the death of a person (mainly young babies and menstruating women), his or her house would be burnt down to escape the harm left by the deceased. This ritual is still followed by some families today. Portrayed in Tsimane' cosmology as a greedy being who selfishly wants to monopolise fish resources for himself, people are in constant fear of disturbing and angering *o'pito* if they kill too many fish and leave them to rot. The migration of fish upriver, for example, is interpreted as a sign of *o'pito* being at work. When this is the case, people refrain from taking a bath or extracting water from the river, all in an attempt not to upset him by disturbing his efforts (Huanca 2006: 157).

Fishing takes place all year round and almost every day. Often, a member of the family goes down to the river in the early morning hours to fish for the day's main meal. When fish is not smoked over the open fire, it can be boiled and made into tasty *jóna*, a kind of thick fish stew that is normally shared with the extended family. The Rio Maniqui or *cojiro*, as it is affectionately called by the Tsimane', serves as the main source of fish, especially during the dry season when low water levels invite the skilled fisherman (and woman) for a good catch. During the rainy season, the Rio Maniqui can become quite dangerous when it overflows its course, flooding the houses and all other vegetation to be found along the riverbank. This is a time when Tsimane' seek smaller rivers, oxbow lakes or creeks for fish. Water dwellers in the Rio Maniqui and its tributaries include caimans, otters, tortoises, river turtles some 86 fish species (Pérez 2001: 56). Most fish are edible, and food taboos are less restrictive as compared to hunted game. Food taboos are generally directed at pregnant women or people who are unwell. Pregnant women, for example, should avoid eating fish with sharp teeth as these may cut the umbilical cord of the unborn baby. Chicchón (1992: 166) contends that the Tsimane' hide from others the fact that they consume stingrays. Stingray liver oil has a high market value and is regularly used as medicine for respiratory infections and their sting equally serves to numb toothache.

The Tsimane' have a wide-ranging armoury to choose from: hooks and lines, nets, arrows, machetes, *barbasco* or fish poisons. Hook and line fishing, though not quite aboriginal to Amerindian societies, has been practiced for a long time and became even more popular once metal hooks became available on the local markets. Formerly, fish hooks were made out of palm wood cuttings and lines plaited from palm fibres. Several hook sizes are used to catch different fish and small fish, insects, pieces of meat or on rare occasions fruits are used as bait. Seemingly enjoying the pleasures of line fishing, an activity almost considered as a form of gathering, children and women set the lines along the river at nightfall and often leave them overnight. While this form of fishing produces rather low results, I still had the impression that the regular fish supply of many households throughout the year truly rests on this seemingly small-time fishing technique. Net fishing, on the other hand, can produce excellent yields if the fishing site is well chosen. Fishing with nets is an exclusively male activity and I have mainly seen it being used in subsiding lakes where fish are somewhat imprisoned, thereby signalling an easy catch. Yet its use is equally rewarding when applied at any promising channel, as the net seals off a little stretch of the water body so that the fish have no chance to escape. Once all fish are captured, the manoeuvre recommences at another section along the waterway.

Contrary to the largely individual fishing techniques described so far, *barbasco* or fish poisoning is a collective undertaking. While mounting concerns about its contaminating effects have contributed to this technique's loss of popularity in the region, it is still practiced on occasion. Men, women and children from different households are all eager to obtain large quantities of fish from this rather elaborate process. It is a skilful technique associated with many Amazonian societies (see Descola 1996; Johnson 2003) that provides not only economic returns, but equally makes for a pleasant family outing and a period for socialising in groups. The Tsimane' use three different kinds of fish poison: *chito'* (*Tiphrocia vogeli*), *conofoto'* (*Hura crepitans*) and *vashi'* and choose certain river bends where water levels are low. Small oxbow lakes may also form a genuine breeding pond for all sorts of fish species and sizes. A temporary dam is constructed using a variety of palm leaves and uva grass poles and the poison, by momentarily changing the chemical equilibrium of the water, makes the fish become *shu'qui* (which literally translates as drunk). By now all participants are ready to catch the fish either with the use of bows and arrows (men and boys), machetes (women and young girls) or simply with their bare hands. Women also scoop small fish up in palm-weaved baskets, keen to take advantage of the fish being intoxicated. When all this is over, the dam is carefully taken apart in order not to impede the further circulation of fish. In terms of labour division, it is interesting to note that the different roles of males and females in fish poisoning somewhat reproduces the gender roles assigned to other modes of interacting with nature. While men are in charge of the heavy and probably more perilous labour tasks (dam construction, preparation of fish poison, the use of bows and arrows), the physically lighter process tasks – to some extent resembling gathering activities – are assigned to women.

As briefly mentioned above, the use of fish poisons has recently caused a wider-ranging debate questioning its ecological sustainability for the river and its resources. Particularly non-indigenous farmers, living near Tsimane' settlements where this fishing technique is still common practice, have raised severe complaints about the contamination of the water body. Also, the park management of the Beni Biosphere Reserve, where some Tsimane' communities are located, has prohibited the application of this technique and started widespread sensitisation campaigns against it. As a result, many communities, especially those more exposed to national culture, have abandoned *barbasco* fishing altogether.

3.3.4 Farming Myths

Along with hunting, gathering and fishing, the Tsimane' are also no newcomers to horticulture. For many centuries, Tsimane' farmers have been cultivating manioc, maize, rice and plantains as their main staples. While a detailed account of the different stages involved in the Tsimane' agricultural cycle will be examined at great length in Chapter 4, in this section I will limit the discussion to the cultural beliefs and perceptions involved in horticultural practices.

For some indigenous societies in the Amazon, horticultural fields are just a place where people work in order to eke out a living (see, for example, Carneiro 1964) and other societies' agricultural practices are almost entirely determined by spiritual beliefs in gardening magic (see, for example, Descola 1996). The Tsimane', so to say, combine a bit of both worldviews. Tsimane' oral traditions surrounding gardening, for instance, are much more limited as compared to the more elaborate world of the forest (see Huanca 1999). Yet both worlds are somehow related, as agricultural practice involves clearing forest patches, which entails the constant negotiation with the *amo*, the spiritual tree master. In former times, special *jemaca* (chants) were sung to accompany the agricultural processes, a veritable art of communicating with guardian spirits that was mostly mastered by shamans.

Huanca (1999) has recollected three oral traditions which he considers to narrate Tsimane' agricultural origins. The first story recounts that in ancient times Tsimane' farmers would mark the boundaries of their field territory with arrows in the shape of a head and wait until a sudden gust of strong wind would topple the trees. The plot was burnt and a small manioc stem thrown onto the soil where it started to grow. What we learn from this myth is that small clearings, caused by naturally occurring strong winds during the winter season, might have been used for cultivation in ancient times. Due to long dry spells following frequent thunder, fallen tree debris would dry much quicker than normally, thereby facilitating the burning process. The second account relates *opoj's* fallow, a legend still heard occasionally in some of the few secluded communities within the Tsimane' territory. *Opoj* is a mythical being, the guardian spirit of agriculture and recognised by the Tsimane' farmer as a plantain-like plant. According to oral tradition, it was *opoj* who introduced the Tsimane' farmer to the art of cultivating plantains. There also exists a different version of this narrative (e.g., Pérez Diez 1983) that associates maize and other cultigens produced by *opoj*. The third reference, though only marginally related to agricultural origins, refers to the tree *puyu* who in the form of a man would work the plots without ever wanting to return home. This is why God transformed him into a tree to remain on the farming site. It is interesting to note that still today, we find many of these tree species in cultivated forest clearings. There also exist myths that recount the origins of the staples manioc and plantain. The manioc myth considers this tuber to have been introduced to the Tsimane' through sky people, who taught the Tsimane' women the art of brewing *shogdye*, a strongly fermented manioc beer. The myth also reveals the power of women over the cultivation of this plant. Still today, the planting, weeding, harvesting and preparation of manioc is largely considered to fall within a woman's domain. Concerning the myth on the origin of plantains, the frugal plantain owner *opoj* would provide the people with bad seeds, forcing them to steal them from elsewhere. This myth illustrates the struggle of Tsimane' farmers to obtain sufficient seeds in the past and equally shows the importance of sharing resources among the people.

From these myths we can deduce that horticultural practice has been around for a long time. It appears that when horticultural practices among the Tsimane' were in their beginnings, agriculture was considered an almost sacred activity. The manioc myth, for example, takes place in the sky and on earth (Huanca 2006: 63). The first maize and manioc harvests were offerings to supernatural beings in order to appease them. These narratives also show that the main staples manioc and plantains have not lost much of their former importance. While it is true that some Tsimane' communities have started to substitute some manioc farming with rice cultivation, manioc beer parties still remain at the heart of the Tsimane' social world. Still today, a meal without plantains – roasted, baked or cooked – is hardly considered a proper meal in most communities. These myths also show that rice, although cultivated in most Tsimane' communities mainly as a cash crop, does not enjoy the same level of cultural importance as compared to the more traditional manioc or plantain staples.

Within the Tsimane' slash-and-burn agricultural system, the cultivation of agricultural fallow plots plays an equally important role. These sites, besides the presence of manioc, maize and plantains, also contain domesticated plants that serve medicinal, material or nutritional purposes. One of these plants is *väij* or the peach palm which is cultivated mostly for its delicious fruits and material importance to manufacture bows and arrows. The ritual surrounding the consumption of the peach palm fruits is quite elaborate and marks the beginning of the year's palm harvesting. The seasonal botanical changes from flowering (September) to fruiting (January), often compared to a woman's pregnancy, go through a number of rituals. When the palm is flowering, women and little children are not allowed to visit the field in order not to upset the guardian caring for the plant. When the fruits are mature, only the oldest man in the family may begin the harvest by following certain rules (e.g., abstention from sexual relations the night before). The first harvest needs to be shared among all extended family members in order for the whole family to feel protected. After the common meal, the family goes to the river to bathe, tossing the peel and seeds into the water as a cleansing process. There, the peach palm ritual comes to an end and from the next day, people are allowed to pick up their daily routines again.

As Tsimane' lifestyles change, so does their belief system. What we see today is a fairly heterogeneous picture of Tsimane' beliefs and practices. As a rule of thumb, Tsimane' communities located at a further distance from market centres have better preserved their traditional spiritual belief system than those living near commercial hubs. Yet in the last four decades or so, people have started to merge their traditional spiritual domains with Christian elements. These developments are rooted in their history of contact with missionary tribes, part of which we are going to explore in the following section.

3.4 Tsimane' History of Contact

3.4.1 From Sporadic Encounter

Tsimane' contacts with outsiders cannot be traced back with absolute certainty, as quite a few historical accounts use generic terms to refer to several indigenous lowland inhabitants. Early references to the groups who inhabited the area today occupied by the Tsimane' include various names such as Rache, Uchumanos, Maniquíes, Amo, Chunchos and Cunana. Yet opinions as to which groups refer to the Tsimane' vary. Whereas Ellis and Araúz (1998) affirm that the Chunchos who lived in the river plains of the Rio Mapiri, Rio Coroico and the Rio Beni are what we call the Tsimane' today, others believe the Uchumanos or Chumanos to be today's Tsimane'.

While opinions diverge on the former denomination of the Tsimane', there is however widespread agreement on the similarities between the Tsimane' and the Mosetene. Both indigenous groups bear a strong resemblance in their language and cultural beliefs and some scholars even consider them as the same indigenous group. While this has not been ascertained, their geographical vicinity may partly explain their similar traits. In 1913 during his travels along the Amazonian rainforest, the Swedish explorer Erland Nordenskiöld considered the Rio Quiquibey to be the frontier between the Tsimane' and the Mosete. When a Tsimane' crosses the river, so he recounts, he used to exclaim: 'I want to know when to return back to my territory' (Nordenskiöld 2001: 172 [1913]). The name Chimane has been used more commonly in literature since the second half of the nineteenth century. Wegner (1931: 87) distinguishes two types of Chimane, the Chimane inhabiting the mountainous areas ('Bergchimanen') and the inhabitants of the floodplains, who he referred to as Churimana ('Pampabewohner'). The Churimana, it seems, no longer exist, whereas the Chimane have largely maintained their traditional way of life due to their isolated lifestyle. Today, the term Tsimane' is commonly used.

The excavation of archeological and paleontological materials, like stone tools, pots and fossil remains in the 1960s, reveals information about the frequent bartering between lowland tribes. Besides these material trading relations, archaeological evidence also shows a permanent flow of cultural exchange between the Tsimane', the Mosetene and other lowland groups through the bartering of tools, medicine and plant material as well as interethnic marriage and language borrowing. Yet trade relations also extended to the highlands. In exchange for forest products such as feathers, monkeys and tropical crops, the Indians of the forest received stone and bronze tools, pottery and highland crops. On an interesting footnote, Saignes (1985) has found similarities in housing construction and belief systems between Lake Titicaca in the Andes and the Rio Beni in the Amazon.

By the late seventeenth century, sporadic encounters with missionaries began. Franciscans, Dominicans and especially Jesuits were all driven by the same goal: 'to subject the barbarian tribes to the Spanish Crown and civilise them' (Pérez Diez 1983: 83). Soon, it was the Jesuits who won the battle over other missionaries and widespread efforts to convert 'savage' indigenous tribes commenced (Reyes-García 2001). After only 7 years of extensive spiritual and intellectual education, the Jesuits established their first mission in 1682, counting some 600 converted indigenous. Soon, they moved up the area from Santa Cruz de la Sierra to Gran Moxos and continued unrelentingly westward to the fringes of the Andean mountain range (Chicchón 1992). In 1693, they established the Mission San Borja along the banks of the Rio Maniqui. For the Jesuits, the location of the town of San Borja was extremely important since it gave them a western stronghold along with missions from other Catholic congregations. Again, whereas most written records do not specify the Indian groups subjugated under the missionary regime, historic sources (see in Ellis and Araúz 1998) did, however, make reference to some 3,000 Churimana people residing in the mission. The missions contributed to the construction of roads and water pipes, the establishment of shops and other small industry and cattle for the common good in all settlements under missionary influence (Lijerón Casanovas 1998). In addition, the priests also tried to appease the Indians by handing out trinkets and tools. Yet, making them stay was by no means an easy task, as most did not at all feel at all obliged to reciprocate by being tied to the mission. According to Jesuit Father Altamirano's written accounts (1699 in Chicchón 1992: 52) on the history of the mission, he considered the Churimana as 'the worst natives of all', an irate remark that may have been due to his inability to dominate them. If our presumptions are true and the Churimana were indeed the ancestors of modern-day Tsimane', then the Tsimane' were already submersed in Jesuit missionary activities since the very beginning.

Approaching the turn to the eighteenth century, uprisings of the Indians against the priests in those missionary settlements became more and more numerous and culminated in setting fire to the buildings and the consequent expulsion of nearly all missionaries (except for the Jesuits) from ethnic territory (Lijerón Casanovas 1998). Other missions disappeared due to the spreading of disease and the resistance of people to living in big communities and adapting to new customs. In 1767, in a struggle for power, the Spanish Crown expelled all Jesuits from their territories in Bolivia and replaced them with other priests, who were mostly Bolivians from Santa Cruz. Due to their lack of virtue and spiritual commitment, it only took a few years for the majority of the missions to vanish from the scene. Since no specific references have been made to the fate of the Churimana, it is presumed that they must have fled back into the forest to settle down along the upper stretches of the Rio Maniqui, where their descendants are still found today.

So despite their continuous and fervent presence in the Moxos area, the Jesuits did not accomplish the conversion of the Tsimane' into a sedentary society dominated by Catholic beliefs. The reason for this to occur, according to some authors, lay in the Tsimane' social organisation. As they lacked centralised leadership patterns, they were somewhat immune to forced attempts to settle them within confined places (Ellis 1996). Others mentioned the limited incentives for the Tsimane' to surrender to the pressures exercised by the Jesuits (see, for example, Huanca 2006). To start with, their subsistence patterns, based on extensive use of resources, prevented them from a sedentary lifestyle in the missions. Secondly, the Tsimane' had learnt to protect themselves from epidemics, previously experienced during contact with white foreigners. Finally, as shamans played a central role as religious leaders for the Tsimane', there was practically no incentive for the Tsimane' to sustain their interest in the missionaries. As we shall see below, two centuries would have to pass for missionaries to return to the Moxos region and rekindle contacts with the Tsimane'. This time, however, on a more permanent basis.

Economic relations with migrants, attracted to the Beni lowlands by the lure of large-scale natural resource extraction, constitute the second set of sporadic encounters between the Tsimane' and outsiders. When Bolivia gained its independence from Spain in 1825, the central political power shifted to La Paz. As a result of these larger political processes, trade relations between the lowlands and the highlands had lost their former importance and when liberal reforms opened the Bolivian market to low-priced European goods, national produce could no longer compete. New ways had to be found to strengthen Bolivia's economic potential and make it more competitive and one attempt was to develop and integrate the remote Bolivian lowland region into a broader national economic strategy. In 1842, the department of Beni was created and economic exploitation strategies based on the extraction of natural resources for the provision of national and international markets fuelled large-sale immigration from the highlands (Reyes-García 2001).

The history of natural resource extraction has been cyclical with its first boom period being the extraction of quinine bark in the north of Beni and the neighbouring department of Pando. From the tree bark, strong alkaloids were extracted which formed the basis for important medicines to control malaria and other fevers. The extraction of quinine had its first peak between 1830 and 1860, and by the year 1860 quinine constituted one of the principal exports to Europe. A second peak period followed between 1877 and 1885 (Roca 1990: 204). Whereas quinine extraction slowly diminished after seeds were secretly exported to Asia, it nevertheless opened the Amazonian routes for the extraction of a second product in demand, namely rubber (*Heavea brasiliensis*). The exportation of rubber from the northern part of the Beni came to be the main pillar of the Bolivian economy along with silver and tin during at least three decades, from 1880 until 1910. The mode of rubber tapping demanded a large work force and a second wave of migration was sparked, this time however, with much greater demographic and social consequences than before. The rubber boom abruptly began to turn into a bust in 1912/13 when the price for rubber on the international market suddenly slumped. Once again, British colonies in Asia began to flood the European market with cheap rubber.[2] What followed was a socio-economic depression and thousands of day labourers lost their jobs. Forced to eke out their living somewhere else and attracted by the vast land resources available in the region, many migrant families moved to nearby town centres.

Through the use of deception and violence, workers were also recruited among native peoples, in particular among the Tacana, Movima and Moxos. Chicchón (1992: 59) asserts that it was an era that had a deep impact on the indigenous populations 'leaving deep scars in native demography, social composition, residence patterns and mythology'. Many authors agree, however, that the Tsimane' were not among the recruited labour force, given their dispersed settlement pattern on the one hand and their marginal integration into the monetary market economy on the other hand. Yet the Tsimane' were still affected by the sudden influx of a new *mestizo* population to the Beni. On the one hand, large-scale migration triggered a process of encroachment on their territory. As a direct consequence, indigenous peoples and newly arrived rubber tappers found themselves competing for the same resources such as game, fish and forest foods. The Tsimane', attracted by the lure of metal tools and other luxury goods, engaged in frequent trading relations with newcomers to their territory. In the end though, encounters between the Tsimane' and migrant populations were still largely characterised by their sporadic nature.

3.4.2 *To Permanent Links*

3.4.2.1 Tsimane' Growing Economic Relations with Outsiders

With many new economic opportunities arising, the Tsimane' history of contact has accelerated enormously over the last century; and with growing exposure, increasingly more permanent links with the outside world have been established. The starting point that triggered economic transformations around the area of San Borja was the sudden influx of hundreds of migrant families who had previously worked for the busted rubber collecting centres in the northern Amazon of Bolivia, to the urban centre of San Borja around 1910. The town of San Borja slowly gained

[2] In1876 the British botanist Henry A. Wickham, extracted 70,000 rubber seeds and smuggled them to England, where they were planted in a glass house in Kew Gardens near London.

economic importance, whereas at the turn of the nineteenth century it was barely more than a little town with a church and the plaza, a couple of retail shops and a few families living here and there (see Huanca 2006: 163). Attracted by the vast stretches of seemingly uninhabited land, new migrant families started to work in agriculture and cattle ranching. Migrants became quite affluent by taking possession of seemingly free land and untamed cattle, yet also by labour exploitation of indigenous people. Labour exploitation involved the practice of debt incursions, meaning that patrons would pay their workers in advance with market goods in order to create huge debt burdens on them and make them dependent and obedient workers. Indigenous people were evidently an easy target as often their lack of Spanish language skills made them vulnerable. In those days, the Tsimane' would occasionally visit San Borja to exchange their forest goods, such as local latex or natural oils for salt, metal tools and clothes. When in town, they frequently interacted with all kinds of people, among them merchants, priests, store owners or housewives. Exchange relations, however, were coloured with discriminatory practices towards the Tsimane' as, in light of their scarcity, commercial goods were preferably sold to white people (see oral testimonies by Huanca 2006: 167).

The construction of San Borja airport in 1936 had a huge impact on the economic development in the region. Connecting the hitherto isolated San Borja area with other commercial centres like Trinidad, Riberalta and La Paz, meant radical transformations of the local economy. While cattle meat, sugar cane and leather exports boosted, daily flights now guaranteed a regular influx of modern commercial goods whose ease of access and higher quality soon became industrial substitutes for the native products such as latex or traditional medicine. At the same time though, the Tsimane', allured by the expansion of retail stores and merchandise arriving at the town, started to travel more frequently to San Borja by river or trail. When marketing of agricultural products began to flourish, so did exploitative labour relations between the newly prosperous white farmers and indigenous peoples in the region. Oral accounts reveal that many Tsimane' were among the day labourers. Many of them got indebted to patrons as a result of which they had to move closer to the town of San Borja. Moreover, the growth of the town of San Borja had another impact on Tsimane' farmers. Some enlarged their farm plot sizes as they now started to produce beyond their subsistence needs to provide for the growing market in San Borja.

In the 1940s, new economic opportunities opened up. These included the extraction of forest products such as quinine bark, a product promoted on a national scale for export to the USA. In exchange for machetes or clothes, the Tsimane' were involved in many ways in the exploitation of this resource. On the one hand, they were contacted for their profound knowledge of the forest. Merchants or contractors themselves needed the Tsimane' to locate quinine trees among the various tree species. On the other hand, various Tsimane' were also directly hired for the extraction of quinine bark. In exchange for low-value tools or clothes, they provided a cheap labour force to contractors, merchants and other indigenous groups like the Yuracaré or Moxeños. At the same time, the Tsimane' were equally valued for their skills to provide forest meat to the groups they were working with. The commercial exploitation of wildlife also took off around the same time. These included the hunting of the black caiman, followed by the commercialisation of animal pelts during the 1950s, 1960s and 1970s when pelts became largely depleted. Again, many Tsimane' were hired as guides and were often given cheap alcohol or a bar of soap in exchange.

From the 1950s, the first river traders started to enter Tsimane' communities to barter commercial products such as fishing hooks, clothes, soap and medicine in exchange for agricultural products such as rice, maize or plantains. As a result, Tsimane' families no longer had to walk to San Borja to obtain market products which they had previously often been denied. Now, as the market came to their communities, more permanent trading relations with the outside world were established. During this period, the Tsimane' became highly skilled in *cajtafa* weaving, a palm that grows at the foothills of the Tsimane' territory in the upriver Maniqui. The Tsimane' weave the leaves into long sheets that serve as high quality roofing material that is highly solicited by people throughout the tropics. With the inception of river trading, the Tsimane' started to use these palm sheets as currency in exchange for market products. After various decades of intensive use, the *caftafa* palm species has become scarce in many parts of the territory. Today, a number of projects are under way in an effort to protect it from further large-scale depletion. By the mid-1970s, logging firms operating in the Department of Santa Cruz started to expand their activities into the Beni region in search of precious hardwoods such as mahagony (*Swetenia macrophylla*), cedar (*Cedrela odorata*) and oak (*Amburana seaerensis*). Independent loggers working under the contract of wood traders needed to rely on the Tsimane' for work, food and orientation in the forest. Tsimane' guides would open forest trails and allow the loggers to cut the trees in exchange for which they were given alcohol or used clothing and trinkets.

3.4.2.2 Colonisation Programmes and Mounting Social Conflict

Despite the growing insertion of the Tsimane' in the regional economy, I believe that it was not until the late 1970s, that the Tsimane' socio-cultural landscape started to undergo changes. These changes, ranging from traditional settlement to subsistence patterns, were largely triggered by two parallel, yet intrinsically interwoven developments: state colonisation programmes in the Tsimane' territory, on the one hand, and the construction of roads, on the other hand.

Following the revolution of 1952, a national land reform was implemented by the Bolivian government in an attempt to encourage colonisation efforts in the lowlands. The idea was to populate the 'empty tropics' and, in so doing, accomplish the economic integration of the Bolivian lowlands. By converting a large part of savannah lands into private property, the Agrarian Reform Law failed to take into account the integral way indigenous people use the land and its resources. Adding to this, the parcelling of land was extremely adverse to indigenous ways of communal land ownership. More so, the new law also declared the State as the owner of the trees and the resources found in the sub-soil, again neglecting the importance of the spiritual elements treasured by indigenous belief systems. Colonisation efforts to the Beni were increased once a commercial road between San Borja and the departmental capital Trinidad was opened in 1975. As the road produced an increase in land values and enhanced the regional export trade, impoverished highland families were impelled to participate in the colonisation programmes. A second and much larger-scale colonisation programme in the area, along the foothills of the Pilón Lajas chain near the towns of Yucumo and Rurrenabaque, took off once the road connecting La Paz with the Bolivian tropics was completed in the mid-1970s. The programme involved the settlement of thousands of highland families, grouped into units of 20 families. Each family received 50 ha of land, usually in the form of square blocks of land spread alongside the road between Yucumo and Rurrenabaque. At hardly any instance during the land distribution process though were the usufruct rights of the Tsimane' population, who had been living there for centuries, taken into account (Robinson 1995). The only attempt undertaken by the Institute of Colonisation was the creation of the 'Tsimane' nucleus', a total of six blocks of land that was allocated to the Tsimane' families living in the area (Huanca 2006: 196). As a result of these efforts, some Tsimane' families began to live in these reduced blocks mingled with colonists, while others sold their land and moved to other, more remote parts of their territory.

Huanca (2006) contends that colonisation efforts, especially along the Pilón Lajas foothills, have triggered a number of social and ecological changes which, in turn, have altered some of the features that determine how the Tsimane' perceive and interact with nature. In social terms, he observes changes in social interaction and lifestyles. When immigrants first arrived, their relationship with the Tsimane' was largely governed by the norm of reciprocity. Food and agricultural labour was shared, both adopting certain behavioural patterns from the other group. Conflicts arose, however, when colonists who lived in the area bordering the Tsimane' wanted to expand their land, either by adding more land to their existing areas or by creating new nuclei. Equipped with legal papers, lawyers, good contacts to town people, the backing of the Colonist Labour National Union and longstanding experience in fighting for their rights, colonists enjoyed many advantages over their Tsimane' neighbours, who all too often lacked the language skills or simply the awareness about existing laws. Yet this is only one side of the coin since not all Tsimane' families were submerged in this unequal power relationship. Huanca (2006) recounts that some Tsimane' families living in the colonisation areas adapted their lifestyles and started to behave like their colonist neighbours. Clearing forest patches for cattle pasturing, a strategy otherwise extremely rare among the Tsimane', became a common economic strategy among these families. Assimilation equally took place in terms of language and Spanish has become the dominant language in the Pilón Lajas foothill area. These socio-cultural changes contrast with the more traditional lifestyles of the Tsimane' living alongside the Rio Maniqui surrounding the town of San Borja, where colonisation efforts have always taken place on a much more limited scale. From an ecological perspective, increasing population numbers have narrowed the availability of land and forest resources in the region. The large-scale influx of colonists has particularly fuelled the clearing of primary forest, since many of them knew how to operate a chainsaw, an experience they have gained in the Alto Beni region during earlier colonisation periods in the 1950s and 1960s. Cut trees have mostly been sold to wealthy urban people for construction or, in fewer cases, to intermediaries for bigger logging firms. The extraction of timber has had a significant impact on the loss of forest resources indigenous peoples exploit for their subsistence. The scarcity of game meat, for example, has been observed by quite a range of anthropologists (see, for example, Chicchón 1992; Daillant 1994). Adding to this is the fact that along the colonisation frontier logging has equally altered the landscape. Turning forests into grasslands means a break in historical and mythical cycles, thereby undermining the socio-ecological relationship the Tsimane' have practiced with their natural environment for centuries. Also, when roads were built, construction companies did not respect ancient Tsimane' hunting trails, archeologically important sites or the habitat of wild animals (Huanca 2006: 160).

In short, it appears that this tense condition is for the most part rooted in the clash of worldviews on land tenure. Like other native Amazonian groups, the Tsimane' have a sense of territoriality which embeds a cultural universe whose frontiers are independent of political borders. As we have learned before, many Tsimane' families still visit historical places like former fishing and hunting areas or visit relatives in other communities. They have a range of myths about places considered sacred to them and identify cosmic markers related to their culture and history. Godelier's (1986: 81–82) definition of territory is emblematic for the Tsimane' territorial concept: 'By territory is meant a portion of nature and therefore space over which a given society lays claims and ensures, for all or some of its members, stable rights of access, control and use regarding all or some of the resources found there, and which it is desirous and capable of exploiting'. For the Tsimane', territoriality does not refer to a permanent occupation of the forest, but a rather temporary and erratic use of its resources along their life cycle (Huanca 2006: 195). In such a way, their communal land ownership regulates access and use rights and, in so doing, presents a stark contrast to the colonists' concept of private property. The following section on Tsimane' land claims has been an institutionalised response to ensuring legal protection measures against growing logging activities by mostly large-scale commercial encroachers.

3.4.2.3 Protective Measures and Tsimane' Land Claims

It was the alarming rate of timber and non-timber resource depletion that called for protective measures. A first attempt to put large-scale timber extraction on hold was the creation of the Tsimane' Forest Reserve by the Bolivian government in 1978. Measuring about 1,200,000 ha, the reserve surrounding the Rio Maniqui covers the majority of the forested area between the towns of San Borja and San Ignacio de Moxos. In the late 1970s, it was home to some 10,000 inhabitants, most of whom were Tsimane', Mojeno, Yurucaré or Movima Indians. In an attempt to further enforce protection measures, by 1982 the government had established an 11% tax on all the wood removed from the Tsimane' Forest (Reyes-García 2001: 8). Yet hopes of full legal protection measures were somewhat short-lived when political and economic pressures mounted to open the reserve to logging companies. Demands were given way to in 1987 when the reserve was opened to seven logging companies and 600,000 ha of Tsimane' Forest was turned into permanent production forest almost immediately. Much has been written about the various control mechanisms to promote sustainable development, but the general picture prevails that timber operators seldom complied with the range of rules and regulations imposed on paper. Adding to this, the forestry service charged with the enforcement of conservation strategies suffered from chronic under-funding and also lacked the technical and human resource capacity needed for effective control (see, for example, Piland 1991).

When granting logging licences to the companies, the government ignored the traditional dependence of the indigenous peoples on the forest. A decree was proposed to title land only on the margins of the Tsimane' Forest, while forestry concessions in the centre of the reserve would be left untouched. Yet after failing to reach a consensus among all the different stakeholders involved, a large-scale protest march was launched to increase political pressure on the national government. More than 700 indigenous peoples representing the Mojenos, Sirionos, Yuracares, Movimas and in fewer numbers the Tsimane' marched for more than 650 km from Trinidad, the departmental capital of the Beni, to the national capital of La Paz. Campaigning for 'Indigenous Rights and Dignity', their strenuous journey took 34 days. Their march captured the interest of the national media and instigated widespread popular support. As a result, in 1990 several decrees for land rights were signed into law. Concerning the Tsimane', communal land titling included an extensive area along the Rio Maniqui and encompassed much of their traditional territory.

The Great Tsimane' Council, an umbrella organisation representing all the Tsimane' people, was founded in March 1989, only months before the indigenous protest march was launched under the coordination of the *Central de los Cabildos de Mojos* (indigenous umbrella council of the Moxos region). The Tsimane' Council entered into an agreement with several logging companies to benefit the Tsimane' by paying a certain percentage of their profits to ensure the functioning of their political arm. After only a few years the project failed and the Tsimane' Council had to withdraw from the contract. Hence, despite their now official rights to the land, timber companies and private contractors have continued their presence in the Tsimane' territory.

3.4.2.4 The Beni Biosphere Reserve (BBR)

The rich biodiversity of the Tsimane' territory has attracted the attention of national and international conservationist organisations and led to the establishment of several conservation projects; the most prominent being the creation of the Beni Biological Station under the auspices of the National Academy of Science in October 1982.

In an attempt to pay tribute to the presence of indigenous people in the reserve, the Biological Station was declared a biosphere reserve by the Man in the Biosphere Programme (MAB UNESCO) in 1986. Biosphere reserves are 'areas of terrestrial and coastal ecosystems promoting solutions to reconcile the conservation of biodiversity with its sustainable use' and are intended to fulfil three complementary functions: conservation, economic development and support in research and monitoring. Within the BBR, where comprehensive management plans have been in place since the mid-1990s, the reserve serves as both, a conservation entity and to a certain extent an indigenous territory. The area of the BBR covers an area of approximately 135,000 ha and is located about 45 km from the foothills of the Andean mountain ranges along the southwest fringes of the Amazon region. It is characterised by a mosaic landscape composed of evergreen seasonal forests, savannahs and swamps (Moraes et al. 2000). Of the total reserve, about 30,000 ha are part of the Tsimane' Territory along the lower Rio Maniqui bordering the community of Campo Bello. In 1988, the BBR rose to fame in the international conservation scene, as it was the first location to participate in the national debt for nature exchange funded by Conservation International. On two occasions, namely in 1997 and 1999, the reserve's limits were adjusted and its boundaries moved further away from the town of San Borja. These changes took place as a result of mounting pressures from influential townspeople who perceived the reserve as a threat to their growing interests in ranching and logging. While the reserve area slightly expanded to the north, the total surface area, however, hardly changed.

The number of Tsimane' families living within the boundaries of the reserve have constantly grown from around 120 households in 1989 to about 180 households in 2000. The majority of the around 20 Tsimane' communities are found on the alluvial terraces along the lower Rio Maniqui and along the upper Rio Curiraba on the south eastern border of the reserve. In addition to these villages we also find various dispersed dwellings of colonist families, ranchers and Tsimane' households scattered throughout the entire reserve. Due to ease of access, itinerant traders tend to visit only the communities located relatively close to San Borja along the lower Rio Maniqui. The general presence of flooded savannahs when penetrating the interior of the reserve makes common access relatively strenuous. This is why some of the remote communities found along the interior stretches of the reserve have preserved a wider range of traditional socio-cultural practices than others.

3.4.2.5 More Recent Missionary Efforts

Missionaries represent another recent group of outsiders with whom the Tsimane' have established growing permanent links. We have learnt before that early missionary attempts in the seventeenth century were characterised by failure. The Tsimane' lived dispersedly along the rivers and the allures offered by missionaries provided not enough incentive to reduce their independent and peripatetic lifestyles to sedentary mission life. Almost two centuries later though, things had changed. In the 1950s, the Tsimane' had already established contacts with outsiders and would frequently interact with merchants in town, loggers, colonists or cattle farmers. River traders would pass by their settlements every so often in search of rice, maize or forest products. Adding to these changes, the national revolution in 1952 set the stage for widespread social reforms which opened up new opportunities for education. While the resources of the national government were taken up by the endorsement of school and educational reforms in the highlands, the state welcomed the idea of delegating the responsibility for schooling the indigenous lowland tribes to missionaries. The missionaries, on the other hand, who now enjoyed the full backing of the national government, saw it as their duty to work with indigenous people, often neglected by the state. To a certain extent, though the missionaries acted on behalf of the state in schooling and health provision, they were still considered more grassroots level actors.

Since the 1950s, two different missionary groups have worked with the Tsimane': the Catholic priests and the Protestant New Tribes Missionaries. Both were primarily concerned with improving health care and schooling programmes among the Tsimane'. Catholic priests started their work with native peoples in the Vicariate of Reyes in 1942. In an effort to strengthen their contacts with the Tsimane', a small group of Catholic missionaries set up a mission in the community of Cara Cara. Yet their first attempt failed since too many outsiders, merchants from the town of San Borja or cattle ranchers, disturbed their plans to protect the uncivilised yet innocent Tsimane' tribe against exploitative forces and defend their rights. As a result, they moved their mission up the Rio Maniqui about 2 years later and changed its name into Misión Fatima. There, much closer to the Tsimane' and largely undisturbed by intruders, the priest had four future aspirations in mind. First and foremost, the mission had to become economically self-sufficient and the Tsimane' should equally benefit from economic development. To this end, 50 ha of forest were cleared and large-scale cultivation of cash crops like cacao and coffee promoted. Furthermore, the priest started to introduce cattle ranching and generated money from the extraction of timber. A second focus was schooling. The missionaries set up a primary school and strongly believed that education would empower the Tsimane' against their white foes. Thirdly, the mission promoted health care and set up a health post for ill people to be treated with modern medicines. Their final and yet most central goal was to indoctrinate the Tsimane'. A chapel was constructed and the Tsimane' were asked to go there every Sunday if they were to accrue health or economic benefits from the mission (Ellis 1996; Huanca 2006). As the mission had to attain economic self-sufficiency within less than 3 years, the priests put particular importance on the promotion of economic development. In the 1960s, the mission bought a tractor, a rice peeler and a diesel generator. It was a novelty in the area and, curious about these new developments, many people moved to Misión Fatima. The community grew enormously and people appreciated the efforts undertaken by the missionaries. Still, attempts to integrate the Tsimane' in the regional market economy remained a localised phenomenon of Misión Fatima and the efforts did not spread to other Tsimane' communities. Things changed with the death of father Martin Baur, a charismatic missionary leader of whom, even today, people speak of with reverence. Following his death, the place has lost much of its former glory and cattle ranching and cash crop farming became largely abandoned. Ellis (1996) asserts that the majority of the Tsimane' families who still live at Misión Fatima or in surrounding communities have largely retained their traditional household economy that constitutes hunting, fishing and the tending small agricultural plots.

On the other hand, the presence of the New Tribes Mission, a group of Protestant missionaries, is today felt much stronger in many aspects of Tsimane' life. The New Tribes Mission is a branch of the Summer Institute of Linguistics, a US based institution that is primarily concerned with the social and spiritual concerns of native peoples. In 1954 they entered into an agreement with the Bolivian government to officially endorse schooling programmes among indigenous peoples in the Bolivian lowlands. When they started to work with the Tsimane' in 1960, their main goals were threefold; to initiate evangelisation by translating the Bible into their native language, to provide educational and health services and to support the Tsimane' in their economic progress. What probably added to their influence in the region was their recognition of the need for bilingual teachers and community leaders to facilitate their defence of communal land rights. The bilingual teacher training was initiated in 1973 and those Tsimane' men interested would receive an intensive 3-month training that equally involved a range of methods on how to teach and preach. In 1984, the New Tribes Mission bought a piece of land on the outskirts of San Borja where it constructed a bilingual training centre, a health post and a radio station. The centre has since provided the infrastructure for short-term vocational training courses on current educational issues, most of which are highly appreciated among the Tsimane' teachers.

Parallel to teacher trainings, New Tribes missionaries have been active in translating the Bible into Tsimane' and providing bilingual didactic material. Huanca (2006: 98) argues that, on the one hand, the translation of the Bible gave them the opportunity to establish permanent contacts with the Tsimane'. On the other hand, the Tsimane' regarded the translation of the Bible as a respectful token towards their culture and native language. To disseminate the word of God in the Tsimane' language, the missionaries introduced portable radios. Today, nearly all extended families own a transistor radio and are eager to enjoy one of the many programmes offered. Tsimane' men are often seen carrying a radio around their neck when embarking on a day's work on the field, and daily conversations are almost always accompanied by the cracking sound emanating from the radio tied to the crossbeams of their dwelling. This is not surprising if we take into account that at present, people can daily choose from around 7 hours of community dialogues, educational programmes and above all, religious messages. Adding to their daily radio presence, missionaries also try to physically visit the communities at least once a year. Their visits are most appreciated by the Tsimane' as they bring along iron tools, fish hooks, clothes or other market articles. They normally stay for 2 days, organise a mass service and attend school classes. At the same time though, they engage in traditional community celebrations. By the late 1980s, the missionaries began to take a firm stand against the growing encroachment policies and exploitative wage labour and trade relations that had been a reality for the Tsimane' for a long time.

In an attempt to organise them collectively, so they could defend their rights, the New Tribes were the driving force behind the creation of the Tsimane' Council (*Gran Consejo Tsimane'*) in March 1989. The Tsimane' Council would be the official body representing all Tsimane', thereby facilitating political negotiations on land claims.

It appears that contemporary missionary efforts have indeed impacted on the Tsimane' ways of life and supported their integration in the regional economy. Yet approaches and impacts between the two groups of missionaries differ. The Catholic priests worked in one community only and focussed primarily on fast economic transformations in an attempt to relieve the Tsimane' from their misery and poverty (Huanca 2006). These attempts brought about changes in the natural environment and introduced the Tsimane' to the monetary economy. The Protestant missionaries, on the other hand, opted for a more inclusive approach on improving education and health care, with less focus on economic progress. Through their large-scale teacher trainings that involve all communities and the design of bilingual educational materials they are more broadly known and, for the most part, respected. Still, Byron (2003) found that missionary efforts have had varying impacts on traditional religious orientations and practices among the Tsimane'. According to her study, communities closer to the town of San Borja have more profoundly altered their religious beliefs than those in remote areas. On an interesting footnote, however, Byron also found that the peoples' approach to this newly acquired religion (people may now consider themselves as either a member of the Catholic or Protestant religion) tends to be rather pragmatic. People do not appear to invest a great deal of their time in catholic or evangelical rites and animistic beliefs and traditional socio-cultural practices are still widely embedded in their daily lives. One of these practices is the drinking of manioc beer, for example. Indulging in regular beer feasts is just one of the pleasures that still permeate Tsimane' community life. For the rest of this chapter, I will look at those features that constitute Tsimane' sociability and community life.

3.5 Settlement and Community Life

3.5.1 Demography and Settlement

As for the Tsimane', a general census on lowland ethnic peoples in Bolivia reveals that the Tsimane' are the fifth largest indigenous group representing 4% of the entire native population of Bolivia (CIDDEBENI 2002). The most recent source on general Tsimane' demography dates back to 2001 and provides an estimate of 8,000 people (Godoy et al. 2005). The earliest information on population trends was provided by Nordenskiöld (2001[1913]). He estimated the total Tsimane' population to range between 2,000 and 3,000 for the year of data collection in 1913.[3]

[3] In his informative travel records he examined one community and concluded that between the years 1897 and 1912 each year 83.5 children in 1,000 people were born – compared to Russia where between 1896 and 1900 the number was 48.4, in Sweden 26.9 (Nordenskiöld 2001[1913]: 176).

Following Reyes-García's (2001) suggestion on Tsimane' growth rates (see Table 3.1.), population grew at an annual rate of 4.86% during 1971–2002. According to these estimates, the population has more than tripled in the last three decades. These growth rates are extremely high compared to annual population growth rates of other indigenous Amazon dwellers.[4] The data provided, however, undoubtedly lacks the reliability and validity that would be necessary for calculating well-grounded demographic trends. First, most figures are merely estimates; secondly, methods for data collection have not been provided in most of the literature stated; thirdly, only within the last decade or so have many Tsimane' obtained identity cards from local municipalities. It may be assumed, therefore, that many individuals were not included in the national census prior to this event.

It is likely that in their course of history the Tsimane' have developed their demographic pattern according to periodic events that required 'adjustments' to such fluctuations. This assumption is based on two notions: periodic flooding and epidemics. Still today, Tsimane' settlements are periodically exposed to river flooding in January or February every year and housing sites are evacuated. I recall a severe flooding incidence in early 2006 when several Tsimane' settlements remained under water for several weeks and people had to relocate their dwellings. As major field crops were destroyed, villagers received emergency food aid immediately following the floods. Epidemic outbreaks are considered the second extreme event that must have brought about demographic adjustments. What we know for certain is that the Tsimane' have been struck by both tropical and Western diseases. Travel histories mention how the Mosetene and Tsimane' were affected by various smallpox outbreaks in 1886, 1894 and 1913. In the 1930s, some Tsimane' communities were affected by measles and no cure had yet been found for adequate treatment. From historical accounts we know that in former times when a person fell seriously ill, the rest of the family would avoid all kinds of contract with the sick person and even move to another place in order to escape from the malicious spirits haunting the ailing individual.

Table 3.1 Population estimates of the Tsimane' according to national census (adapted from Reyes-García 2001)

Population	Year	Source
2,000–3,000	1913	Nordenskiöld (2001)
1,800	1971	Kelm (1972)
2,000	1974	Riester (1978)
5,000	1989	Comisión Socio-Económica (1989)
5,694	1994	CIDDEBENI (1999)
7,385	1999	CIDDEBENI (2002)
8,000	2001	Godoy et al. (2005)

[4]The Matsigenka of Peru, for example, have an annual growth rate of 2.30–3.19%.

Gurven et al. (2007) examined the regional and temporal trends in mortality patterns among the Tsimane' population between 1950 and 2000 and found that villages in the remote forest regions show the highest overall mortality, while half of all deaths are due to infectious disease, especially respiratory and gastrointestinal infections. Gender differences in mortality are mainly evident during reproductive adult years and late adulthood. Women in fact show a 35% higher mortality from 16 to 39, while men demonstrate a 50% higher mortality after 60. Riester (1976) contends that the Tsimane' practice infanticide on children who are born deformed, a common practice also found in small tribal communities.[5] Gurven et al. (2007: 388) somewhat defuse this assertion, emphasising the relatively low numbers of infanticide among the Tsimane' when compared to reported numbers for other tribal people in the Amazon region. According to their long-term analysis, infanticide accounted for merely 5.3% of all infant deaths between 1950 and 2000.[6]

Albeit somewhat loosely, we find Tsimane' demography to be linked to the formation of settlements. While early historical accounts have come up with different conclusions concerning the stability of early Tsimane' settlement patterns,[7] what we have sure evidence of is that their choice of settlement would largely be dictated by the geographical vicinity of a river. Traditionally, most hamlets were formed on a temporary basis and rarely exceed 5 to 6 households related by blood; a measure that prevents the overexploitation of game and fish resources. This strategy has also been common with other Amazonian peoples; the Matsigenka of the Peruvian Amazon, for instance, are known to move to a new homestead every 4 years (Johnson 2003). On a similar note, Descola's (1996) study on the Achuar equally argues that the rapid depletion of important game resources is the main trigger for the peoples' prevalent 'taste for movement'. Today, the Tsimane' are becoming increasingly sedentary. Growing missionary influences, increased exposure to the market economy and the introduction of primary schools in many communities have all instilled a sense of community among the Tsimane'. These developments have created opportunities for integration and leadership that would otherwise not have taken place. Formal community structures facilitate the transactions needed to negotiate and obtain legal land titles and, in so doing, combat the growing intrusion of cattle ranchers, colonists or loggers into Tsimane' territory. Changes in Tsimane' settlement patterns, however, have occurred to different degrees. Today, for example, only 60% of their villages have primary schools (Godoy et al. 2005: 143), while others,

[5] See, for example, the Bushmen of the Kalahari Desert (Lee 1979) and the Matsigenka in the Peruvian Amazon (Johnson 2003).

[6] Gurven et al. (2007: 388) further reported that the main causes for an infant's pre-natal death are due to the mother's traumatic fall (36.5%), from over-working or carrying too much weight (16%) and maternal sickness (15%). An additional 7% was self-induced as a means of spacing births or because of doubts concerning paternity.

[7] While some early historical accounts characterize the Tsimane' as a semi-nomadic tribe (Metraux 1948; Riester 1976), others emphasise that many Tsimane' settlements have geographically stable histories. Hissink (1955) and Hissink and Hahn (1952) on their adventurous travels along the Rio Maniqui describe settlement clusters which can still be located in the same places today.

mostly smaller and more remote settlements still retain their traditional social structures and are practically invisible on official maps.

A common feature for most Tsimane' families are their great mobility. The Tsimane' are enormously mobile and often move seasonally to take advantage of fluctuations in natural resources. Yet despite their great mobility it is also interesting to note that most of the moves they make are within their own ethnic territory and within particular boundaries for each extended family cluster. During certain periods of the year, families may move to river beaches and exploit the abundance of fishing resources. At other times when game animals are fattest, it may well be the forest that becomes a more solicited temporary dwelling place.

3.5.2 *The Private and the Public Sphere*

To the outside observer, the everyday social world of the Tsimane' seems to strike a balance between periods of dispersed nuclear family living, aggregation in extended family networks, and community life. Yet it is the household that provides the heart of Tsimane' sociability. Though providing for a single unit, the household is strongly embedded in a larger social setting that includes extended family networks as well as an attenuated network of friends, neighbours and other actors constituting life in the community.

3.5.2.1 The Household

The Tsimane' household provides for the main unit for production, consumption and social reproduction. People preferably live in nuclear family dwellings, spaced well away from their neighbours' homesteads. I was told by my Tsimane' hosts that first and foremost, the vicinity to kin members provides for the main parameter when selecting a housing site. What comes next is the need for a good and reliable source of water, followed by the presence of fertile soil. Though most Tsimane' households are built along the Rio Maniqui, they are not directly built on the river bank but a further distance away. This is mainly for two reasons: first, fluctuations in water level from wet to dry seasons bring the inherent danger of flooding, as happens periodically. The second reason is no less important, since maintaining a short distance away from the river makes it just about possible to abide the persistent presence of mosquitoes and sand flies that infest the riverbanks, particularly during the wet season. A typical Tsimane' dwelling impresses the outsider by its apparent sturdiness. The traditional Tsimane' house is rectangular, about 6 m long and 4–5 m wide, with a palm-leaf roof that peaks about 4–5 m off the ground. Dwelling huts in more remote communities are still found without any walls or simply one palm-wood wall facing south for protection against the strong *surazo* winds sweeping the area in May and June every year. But even there, more modern constructions comprising four walls and a single doorway are becoming increasingly popular. On first entering from bright daylight, a walled house can be quite a dark and smoky place to be. Each household has its own clearing that serves as a patio and a separate kitchen building that is located near the main dwelling.

From an early age girls and boys are socialised into performing the duties that they will be expected to carry out as an adult. Girls usually help their mothers in cooking, fetching water and firewood, sewing and weaving and taking care of siblings, whereas boys can be observed playing with bows and arrows, catching small fish at the riverbanks and often accompany their fathers on hunting trips. Both girls and boys are raised to work in the fields, performing tasks such as sowing rice or weeding. In communities where there are schools, usually boys are more encouraged to attend than girls, as they often have to look after their siblings at home. Husband and wife form the centre of production and reproduction and are quite independent from other households. They are free to decide when to leave the hamlet and wander off into the forest or on a fishing trip. A Tsimane' marriage may be considered a partnership of two equally skilled individuals, each with a separate sphere of influence, yet whose different roles underline their importance as a complementary pair in everyday production. Their socialised behaviour tells a story of clear differentiation, one that leads not to segregation and discrimination between the sexes, but to mutual respect and interdependence.

Respect for personal space within the household is also expressed in the general pattern of property rights. All kinds of household resources have, in fact, distinct owners. Ownership is created through the use of labour and can only be transferred through barter or gifting. It is a social rule that is still widely respected among the Tsimane'. One incident brought this home to me upon my return to the village of Campo Bello after an absence of 1 year. When inspecting my former dwelling, Andrea proudly showed me the banana tree I had planted when I first arrived to the community. She pointed to the tree and described meticulously how she had cared for it while I was away and that she was happy to now hand it back to me. Everything that belongs to the household is individually owned. The house belongs to the man who built it and people do not commonly enter the house when family members are away. The notion of ownership begins young and from an early age, every child is socialised into keeping track of his or her possessions. When a child is asked to name all his or her belongings, a comprehensive list of all the items ranging from pencils to pets is quickly provided. Manufactured items like bows or arrows equally contain small signs that give hints of the producer. People can easily identify the items fabricated by them and generally remember who gave them the items they own or, in case of items obtained at the market, where, when and from whom these commodities were obtained.

3.5.2.2 Beyond the Household

Every nuclear family is embedded in a larger collective. What traditionally used to be the extended family network only, is the larger community setting today, that is increasingly becoming important. Since every kind of outside interaction goes hand in hand with certain sets of behaviour rules to be followed, I would firstly like to inquire into the reproduction of kinship relations before moving the discussion to life in the larger community setting.

Kinship and kinship ties are reproduced through marriage. The Tsimane' traditionally followed a Dravidian kinship system of cross-cousin marriages which centred on the extended family and formed the basis for social organisation (Chicchón 1992; Daillant 1994; Ellis 1996). Today, polygyny and other traditional forms of prescribed marriage have for the most part been eroded and given way to monogamous relationships. Especially in larger settlements under strong missionary influence, traditional practises have widely been discouraged and even prohibited. According to Tomás Huanca (2004, pers. commun.), polygynous marriage patterns still exist in very remote communities and account for nearly 3% of Tsimane' families. Tsimane' kinship terminology prescribes marriage with a person falling into the category of *fom'* (cross-cousin) and still today, many Tsimane' regard this category to be the only correct marriage. In Ellis' (1996: 80) view, choosing a 'correct' marriage partner serves two purposes. First, marrying a close relative is to avoid the unknown attributes of powers of sorcery and anger attached to people unknown. Secondly, this kind of marriage enables an enriching of the kindred network, often across large distances. Tsimane' marriage patterns though have undergone changes in recent years. In San Antonio, for example, a large Tsimane' community close to San Borja, I have observed a growing number of exogamous unions between Tsimane' and Yuracaré, Movima and other non-Tsimane'.

A woman is ready for marriage after her first menstrual period and a man usually looks for a potential bride once he has established a reputation in hunting, fishing and farming. When contemplating marriage, both men and women particularly look for someone who is hard-working. It would be a grave error to marry someone who is lazy as such a spouse will fail to hold up his or her end of the division of labour. Among the possible spouses, an attractive wife is defined as much by her expert skills in the household or economic reproductive sphere (e.g., being an expert weaver) as well as by conformation to the indigenous canon of physical beauty. For a woman, on the other hand, a good husband is one who satisfies the biological needs of his wife (by fulfilling his sexual and hunting obligations). Only when these expectations are fulfilled, Tsimane' adolescents are ready for marriage. When a man and a woman court, their families vigilantly cross-examine their prospective in-laws. If the marriage brings together a man and a woman from different communities, the common post marital residence pattern is for the man to move into the bride's home and work for her father during a year or two.[8] Following Matteson's (1954: 79) description of the indigenous Piro, 'matrilocal residence patterns are a matter of protection for the bride and of mutual help'. Indeed, if a marriage fails during this time, the newly-wed husband may simply leave; a move that equals

[8]Ellis (1996: 52), however, warns about establishing strict rules of rigidity on Tsimane' residence decisions, as she has various case histories that simply do not comply with such patterns.

divorce. After the trial period, typically marked by the birth of the first child, the young couple has the freedom to choose where to settle next and establish their own household unit.

A glimpse on the housing topography within the boundaries of a larger Tsimane' community reveals that extended families repeatedly live in their own clusters. Like stones in a mosaic, this collection of individual huts tends to retain its shape even when integrated into a larger community. Migration histories document that households tend to remain in the general vicinity of a group of families over time. Those smaller units usually consist of siblings or cross-cousins that have remained closely linked through intermarriage. The close social ties between extended family networks are usually expressed through the sharing of food and other valuable items. Ellis (1996) refers to the extended family as a kind of social net to ensure that all family members are looked after. The fact that Maberto's wife had left him, for example, did not mean abstinence from well fermented manioc beer as he was allowed to freely participate in beer drinking sessions at his brother's. The extended family network equally allows for a more flexible distribution of skills. If a man's favourite pastime is hunting, he is welcome to engage in frequent hunting trips and share it equally among extended kindred upon his arrival back home. In return, someone else may invest more time to look after the manioc or rice plants, thereby optimising the overall time investments to ensure a varied diet for all members. Yet despite certain intimacies shared by extended kindred, individual households still take measures to maintain some level of privacy. Individual dwellings, for example, are repeatedly found a small distance away from each other, allowing a row of plantains or fruit trees to separate them and ensure some possibilities for retreat. Also, a number of households clear the individual trails leading to their huts so as to ensure that people heading on the main path need not pass any given house unless it serves as their destination. When newly-wed Roque built his house within the environs of his parents' patio, he made sure that the entrance was placed well away from his parents' entrance, facing the river. It was a measure to protect their still young intimacy and to come and go at the pace the young couple would decide.

Life in the community provides for the second sphere of social living. Interestingly, in terms of daily time use, which will be discussed in detail in Chapter 5, the Tsimane' only spend a small fraction of their time resources in communal settings. These periods of interaction, however, are undeniably significant for social reproduction and are generally indulged in with joy. Communal work provides the first such social gathering. It has become increasingly important and is mostly considered a male task. Clearing communal pathways, for example, requires a certain date to be set by the village leader and people gather outside the school building. It seems that these tasks are taken lightly and when moving slowly forward along the forest path, it almost appears as a social event where men fire off jokes. When tired from mowing and weeding, a communally prepared meal by some of the local women awaits the hungry men. The meal rounds off the few hours of group interaction, during which strenuous physical work is generally combined with playful sociability. In school communities, communal work efforts also comprise the regular cleaning and maintenance of the school building. Just like in our societies,

Tsimane' parents are worried about a leaking roof or a brittle wooden post that might endanger the lives of their children attending classes.

A central feature of community life is the sharing of communal beer feasts. Sharing *shogdye'*, strong homemade manioc beer, with other community members forms the core of Tsimane' sociability and strengthens their ethnic solidarity. The social relations created in beer distribution reach beyond the confines of the household and include co-residents and even *napo'*, non-Tsimnane', visitors. As many Tsimane' candidly express, there is little point in visiting close kin or neighbours in the absence of beer. Beer is always offered to passers-by, if present, and its consumption creates a somewhat affective atmosphere. The production of beer is essentially a female domain and takes two or three women a whole afternoon to prepare. Sweet manioc is the main ingredient of most kinds of beer and depending on the recipe, may be combined with maize and/or ripened or slightly sour plantains. At times, beer is made from maize alone but it is the perfect mixture of maize and manioc that is believed to ferment into the most enjoyable beer necessary for the feast to be a success.

3.5.2.3 Patterns of Leadership

The Tsimane' are a perfect synthesis of those enigmatic inclinations peculiar to many societies of Amazon Indians. 'Egalitarian' in any typology of comparative political systems, they have until recently lacked institutions and democratically elected community leaders. The concept of leadership was largely absent in traditional Tsimane' society and, due to their healing powers and their ability to communicate with all kinds of natural spirits, only shamans (*cocojsi'*) would have a higher hierarchical standing within the larger social setting. When arguments arose, people would frequently resort to a shaman, hoping to settle the dispute and heal the people from anger or distrust. Their healing powers were commonly believed to be supernatural and still today individuals seek the advice of a shaman when feeling unwell. Nonetheless, the role of shamans is increasingly losing importance in the contemporary Tsimane' world setting as new leadership roles are created. On the community level, increasing exposure to national society has urged the Tsimane' to elect community leaders. Based on a common system used in national communities, there are three authority roles: the *corregidor* or village chief, the *alcalde* or mayor and the *presidente de junta escolar* or president of school union. Ideally, all three representatives should be familiar with the national language, Spanish, as they act as mediators between their communities and the outside world. Likewise in school communities, teachers repeatedly enjoy a privileged status and authority over other community members, often based on their educational accomplishments as well as their Spanish language abilities.

Finally, the respect individuals evoke is not static but rises and falls within a person's life cycle. Seniority, for example, invokes a higher social standing as compared to younger members of the community. In Campo Bello, Angel's house was considered the social centre of the extended Tayo family cluster. His children would pay regular visits to him and social gatherings mostly took place on his patio. On a similar note, community members who excel in one productive activity or another tend to enjoy an overall higher social standing than others. Others are known for their sharp wit, astuteness or wisdom. All these attributes equally infer a certain sense of social acceptance and status.

3.6 Conclusion

This chapter examined the various angles from which to look at the different dimensions that constitute the Tsimane' world. It opened up with a general description of the biogeographical features of the Tsimane' territory. Their natural habitat harbours a variety of tree species, game animals and soils. Followed by a discourse on the cultural appropriation of natural resources, I have described the ways of using nature in terms of spatial segmentation – forest, river and the farming site. The latter, however, solely focussed on the cultural beliefs of horticultural practices, as the stages of the Tsimane' agricultural cycle and the labour involved will be treated in the following chapter. In their worldview, people interacting with nature need to show respect for the natural resource they utilise in order not to upset its guardians. The Tsimane' daily routine therefore combines the constant need for material satisfaction with the fear of being bewitched. The discussion then shifted to trace the peoples' history of contact, from sporadic encounters with outsiders to gradually more permanent links with the regional market economy. In the last section of this chapter I have given a short account of the social setting that forms part of everyday life. We can distinguish between the social topography within the confines of the household where the married couple comprise the centre for productive and reproductive activities, each with their own and equally important sphere of influence. Beyond the privacy of the household, people are well aware of the sets of behavioural rules governing their interaction with others. Extended family networks ensure food security and in the absence of a spouse, provide the necessary social net for individual relatives. Socialising in groups is mostly enjoyed when accompanied by manioc beer; while only a small fraction of time is actually used in communal activities, these periods of interaction are increasingly cherished by community residents. While it is hoped that the first two chapters have given the necessary theoretical and empirical setting, we can now return to the permeating thread of this monograph: the empirical analysis of the environmental relations the small Tsimane' community of Campo Bello maintains. It is this interaction at the society–nature interface we are now about to explore.

References

Apaza, L., Wilkie, D., Byron, E., Huanca, T., Leonard, W., Pérez, E., et al. (2002). Meat prices influence the consumption of wildlife by the Tsimane' Amerindians of Bolivia. *Oryx, 36*(4), 382–388.

Byron, E. M. (2003). *Market integration and health: The impact of markets and acculturation of the self-perceived morbidity, diet, and nutritional status of the Tsimane' Amerindians of lowland Bolivia.* PhD thesis, University of Florida.

Carneiro, R. (1964). Shifting cultivation among the Amahnuaca of eastern Peru. *Völkerkundliche Abhandlungen, 1*, 1–18.

Chicchón, A. (1992). *Chimane resource use and market involvement in the Beni biosphere reserve, Bolivia.* PhD thesis, University of Florida.

CIDDEBENI. (1999). *Plan de gestión territorial indígena de la TCO Chimane (fase diagnóstico).* Bolivia: San Borja.

CIDDEBENI. (2002). *Diagnóstico del distrito municipal indígena Chimane, proyecto: Apoyo a la formulación del plan distrital de desarrollo indígena.* Trinidad: CIDDEBENI.

Daillant, I. (1994). *Sens dessus dessous: Organisation sociale et spatiale des Chimane d'Amazonie bolivienne.* PhD thesis, Laboratoire d' Ethnologie et de Sociologie Comparative, University of Paris.

Descola, P. (1996). *In the society of nature: A native ecology in Amazonia.* Cambridge: Cambridge University Press.

Ellis, R. (1996). *A taste for movement: An exploration of the social ethics of the Tsimanes of lowland Bolivia.* PhD thesis, University of St. Andrews.

Ellis, R., & Araúz, G. (1998). Planificación, PNUD: Programa Indígena: La pay Pueblo Indigena Tsimane'. In J. M. Caller (Ed.), *Pueblos Indígenas y Originarios de Bolivia* Ministerio de Desarrollo Sostenible (pp. 1–23).

FAO (1979). *Sandy soils.* Vol. 25.

Fischer-Kowalski, M., & Haberl, H. (2007). Conceptualizing, observing and comaring socioecological transitions. In M. Fischer-Kowalski, H. Haberl (Eds.), *Socioecological transitions and global change: Trajectories of social metabolism and land use* (pp. 13–62). Cheltenham, UK, Northampton, USA: Edward Elgar.

Godelier, M. (1986). *The mental and the material: Thought economy and society.* London: Blackwell Verso.

Godoy, R., Reyes-Garcia, V., Huanca, T., Leonard, W. R., Vadez, V., Valdés-Garcia, C., et al. (2005). Why do subsistence-level people join the market economy? Testing hypotheses of push and pull determinants in Bolivian Amazonia. *Journal of Anthropological Research, 61*, 157–178.

Gurven, M., Kaplan, H., & Zelada Supa, A. (2007). Mortality experience of Tsimane Amerindians of Bolivia: Regional variation and temporal trends. *American Journal of Human Biology, 19*, 376–398.

Hanagarth, W., & Sarmiento, J. (1990). Reporte preliminar sobre la geoecología de la sabana de espíritu y sus alrededores (Llanos de Moxos, Departamento del Beni, Bolivia). *Ecología en Bolivia, 16*, 47–75.

Hissink, K. (1955). Felsbilder und Salz der Chimane Indianer. *Paeduma, 6*, 60–68.

Hissink, K., & Hahn, A. (1952). *Chimane. Notizen und Zeichnungen aus Nordost-Bolivien.* Stuttgart: Franz Steiner Verlag Wiesbaden GmbH.

Huanca, T. (1999). *Tsimane' indigenous knowledge. Swidden fallow management and conservation.* PhD thesis, University of Florida.

Huanca, T. (2006). *Tsimane' oral tradition, landscape and identity in tropical forest.* La Paz: Wa-Gui.

Johnson, A. (2003). *Families of the forest: The Matsigenka Indians of the Peruvian Amazon.* Berkley: University of California Press.

Kelm, H. (1972). Linden-Museum. *Eine Restgruppe der Südost-Pano im Oriente Boliviens.* Stuttgart, Tribus 21.

Lee, R. B. (1979). *The !King San: Men, women and work in a foraging society.* London: Cambridge University Press.

Lijerón Casanovas, A. (1998). *Mojos-Beni: Introducción a la historia amazónica.* Trinidad: CIDDEBENI.

Matteson, E. (1954). *The Piro of the Urubamba.* Paper 10, Berkeley: Kroeber Anthropological Society, University of California.

Mayer Roca, F., Caymani, C. C., Mayto, M., & Huanca, T. (2000). Como es la cacería y nuestros animales, chiyacseja' jebacdye'. *Publicaciones Proyecto de Investigación No. 65*, CIDOB-DFID, Santa Cruz: CIDOB.

Metraux, A. (1948). The Yuracare, Mosetene and Chimane. In J. Steward (Ed.), *Handbook of South American Indians*. Washington: Smithsonian Institution, Bureau of American Ethnology.

Moraes, R., Ribera A., & E. Villanueva (2000). La vegetación de la estación biológica del Beni y su importancia para la conservación. *SI/MAB Series No. 4*, Washington, D.C.

Nordenskiöld, E. (2001). *[1913] Exploraciones y aventuras en Sudamérica*. APCOB: La Paz.

Odum, E. P. (1973). *Fundamentals of ecology* (3rd ed.). Philadelphia: Saunders.

Pérez, E. (2001). *Uso de la ictiofauna entre los Tsimane*. Universidad Nacional Mayor de San Andrés: B.S. thesis.

Pérez Diez, A. (1983). *Etnografía de los Chimane del oriente boliviano*. PhD thesis, University of Buenos Aires.

Piland, R. A. (1991). *Traditional Chimane agriculture and its relation to soils of the Beni biosphere reserve, Bolivia*. University of Florida: Master thesis.

Reyes-García, V. (2001). *Indigenous people, ethno-botanical knowledge and market economy: A case-study of the Tsimane' Amerindians in lowland Bolivia*. PhD thesis, University of Florida.

Richards, P. (1985). *Indigenous agriculture revolution: Ecology and food production in West Africa*. London: Hutchinson.

Riester, J. (1976). *En busca de la loma santa*. La Paz: Los Amigos del Libro.

Riester, J. (1978). *Canción y producción en la vida de un pueblo indígena: Los Chimane, tribu de la selva oriental*. La Paz: Los Amigos del Libro.

Robinson, G. (1995). *Comparación del impacto relativo de sistemas de chaqueo y algunas alternativas cerca a Yucumo*. Beni: Ruralter.

Roca, J. L. (1990). *Economía y sociedad en el oriente Boliviano (Siglos XVI–XX)*. Santa Cruz: Editorial Oriente S.A.

Roosevelt, A. (1989). *Resource management in Amazonia before the conquest: Beyond ethnographic projection, in resource management*. New York: The New York Botanical Garden.

Saignes, T. (1985). *Los Andes orientales: Historia de un olvido*. Cochabamba: CERES/IFEA.

Sanchez, P. (1976). *Properties and management of soils in the tropics*. New York: Wiley-Interscience.

Socio-Económica, C. (1989). *Bosque Chimanes: Estudio socio-económico*. La Paz: Ministerio de la Presencia.

Tobías, A. (1998). *Relaciones económicas y sociales de la etnia Chimán con la sociedad regional*. Master thesis, Sociología UMSA, La Paz.

Wegner, R. N. (1931). *Kosmos: Handweiser für Naturfreunde*. Stuttgart: Frankfurter Verlagshandlung.

Chapter 4
Stocks, Flows and Land Use: The Metabolic Profile of Campo Bello

Abstract This chapter offers the empirical results of a local material, energy and land use analysis in the village community of Campo Bello in the Bolivian Amazon. It opens up with a general village description and a discussion of local demographic trends. Against this backdrop, the local metabolic study will be discussed along the following lines: (1) the data collection process, (2) the presentation of the metabolic results and (3) a biophysical discussion on the future prospects of the community. As far as the first is concerned, I am going to illustrate the data collection process in the field, pointing to the hurdles and challenges encountered during the process. The next section entails the calculation of biophysical indicators. This comprises the analysis of all biophysical stocks and flows (in terms of quantity, weight and nutritional values) that enter the social system, get transformed, and leave again. The biophysical analysis then moves to describe the local system's land use practices and its human appropriation of net primary energy (HANPP).

4.1 Introduction

We are now to empirically explore the biophysical landscape of Campo Bello, a small indigenous community in the Bolivian Amazon. This journey will provide us with a comprehensive picture of the production intensity and environmental pressures the peoples' use of resources (material, energy and land) entails. The operational concepts applied for this purpose are the notions of social metabolism, on the one hand, and colonization of terrestrial ecosystems, on the other hand. Although they are interrelated, each concept also has its own purpose; while the first is concerned with the calculation of material and energy flows, the second calculates the impact of land use.

This chapter starts with a general site description of Campo Bello, followed by an analysis of changing local demographic trends. I then focus on the political economy of the local community, thereby hoping to provide a better understanding of the system-specific process dynamics at the society–nature interface.

L. Ringhofer, *Fishing, Foraging and Farming in the Bolivian Amazon: On a Local Society in Transition*, DOI 10.1007/978-90-481-3487-8_4,

Against this backdrop, the subsequent metabolic study provides the heart of this chapter and constitutes three interrelated sections: (1) the data collection process, (2) the presentation of the metabolic results and (3) a biophysical discussion on the future prospects of the community. As far as the first is concerned, I am going to illustrate the data collection process in the field, pointing to the hurdles and challenges encountered during the process. The next section entails the calculation of biophysical indicators. Within the realm of the MEFA framework, I am going to apply the first set of relations: the concept of social metabolism. To this end, all biophysical stocks and flows (in terms of quantity, weight and nutritional values) that enter the social system, get transformed and leave again, are traced and sustainability indicators established accordingly. What then follows is the application of the second set of relations in order to analyse the impact of land use practices: the colonization of terrestrial ecosystems. Since agriculture presents the main socio-economic activity that implies a deliberate intervention into the natural system, this chapter moves on to give a lengthy description of the agricultural production systems prevalent in the community. The calculation of HANPP provides us with an indicator to measure the colonization intensity caused by the system's specific land use practices. Finally, the different strands are woven together into a biophysical discussion that, in light of a sustainability transition, highlights likely and unlikely futures for the local community of Campo Bello.

4.2 The Community Setting

Campo Bello is just one of many Tsimane' villages scattered along the meandering course of the lower Rio Maniqui. Officially granted community status in April 1995, the villagers live in single-family dwellings within extended family clusters, often spaced well away from their neighbour's houses. According to the records kept by the village headman, the community first became inhabited in 1910 and judging from narratives given by a few elderly Tsimane' informants, we can deduce that in the beginning the community comprised two extended families only, the Tayos and the Durvanos. In former times, the community used to be called *Chomindya' ya'* in reference to a palm species that grew in abundance in those days. Today, however, this plant has lost its former importance and is no longer found close-by. The area within the village boundaries of Campo Bello is covered by evergreen forest that spreads like a blanket alongside the river. Moving beyond the forest stretches, away from the watercourse, we find humid and dry savannah formations. The humid savannah is seasonally flooded and due to its loamy texture not suitable for agricultural usage. The fauna found in the savannah region is adapted to seasonally flooded soils and comprises monkeys, ant eaters, deer as well as various reptiles near the water holes. The community lies about a day's canoe journey downriver from the market town of San Borja.

While GPS data measures a linear distance of around 13 km from San Borja town centre, the distance calculated by waterway amounts to 23.7 km from the port of San Borja (GTZ 1997). Campo Bello can either be accessed by canoe or motor-driven boats, by moped, bicycle or simply on foot. As access by land may be an arduous affair when rains are abundant, the river remains the only mode of transport during this period.

41 households are dispersed along both banks of the river for a length of around 3.5 km (see Fig. 4.3). As observed in most other Tsimane' settlements, all dwellings are built in the immediate vicinity of the river. Since the people are aware of the constant possibility of river floods during the rainy season, residential huts are mostly found on elevated alluvial terraces. Travelling the course upriver for about an hour on foot (or half an hour by canoe) we come across the community of San Antonio, a slightly larger settlement comprising of Tsimane' and residents from three other ethnic groups, the Yuracaré. Movima and Mosetene. Travelling down-river, an almost invisible foot trail that curls its way through the forest coppice, takes the explorer on a 3-h walk to Monte Rosa, a small hamlet that consists of merely a few extended family households.

Figure 4.3 reveals a higher population density along the eastern river bank, comprising 23 households as compared to the 18 households found on the western edge of the river. This may possibly be a result of the soil qualities found in the area. Piland (1991) affirms that the eastern river bank along the entire area down-river of San Borja generally contains higher soil fertility. Interestingly though, this view is not unanimously shared with Campo Bello residents. On several occasions, informants mentioned the high chance of encountering *jak tsincus* along the western bank, a rare black type of earth that, because of its fertile dark humus surface, enjoys a good reputation for rice production. What remains uncontested, however, is the fact that residents on the western bank push up against infertile savannah lands that are used for cattle ranching and serve as communal hunting grounds. These householders are thus under greater land pressure than residents on the more populated eastern bank who are surrounded by lush primary and secondary forest areas. At the time of research in 2004/2005, two families (households 16 and 32) had taken up short-term secondary residence on the opposite river bank and both had opened new agricultural plots near their newly established temporary dwellings.

The school, to a certain extent, represents the centre of the community and acts as a magnet for many families who do not to settle too far away from it. For the children from the household farthest downstream (household 1), it is a 25 min journey to travel from their homestead to school. The school premises and their adjacent soccer field are also the place where every day communal interaction and the occasional village feast take place. Engaging in playful sociability, men, boys and girls frequently play soccer or other games with cries of fervour and are fiercely competitive with other communities. Likewise, the families of Campo Bello are no exception to other Tsimane' communities and levels of excitement rise as the day for a village celebration approaches. This has not always been the case, as their newly emerging 'sense of community' is in stark contrast to many of their historic preferences about where and how to 'best' live. In former times, Tsimane' families were inclined to avoid dense settlement patterns for two reasons. First, it made them more vulnerable to epidemic outbreaks and secondly, with rising population density, the local resource base was said to dwindle all too rapidly.

Over the past decades, the village has increasingly become exposed to outside influences. On the one hand, New Tribe missionaries have introduced a Wednesday and Sunday Mass service and trained a couple of the younger community members to say Mass. In return for their loyalty, residents are given medicine and other useful commodities during the missionaries' visits in March every year. On the other hand, the village has also witnessed the implementation of a number of development projects administered by local authorities and non-governmental agencies. As to the former, development efforts include the construction of a concrete school building in 1993, the installation of electricity for the school building and a solar panel for the operation of a communal telephone.[1] In 2001, a development project installed several individual latrines and concrete wells. In 2006, a new project arrived to the village. Still operational, it involves various families in the cultivation of beans and the raising of poultry. After a severe river flooding in 2006, the village also benefited from, albeit sporadic, emergency food aid by the World Food Programme that was ongoing for several months.

Compared to other Tsimane' communities, conflicts with encroachers are much less pronounced in Campo Bello. Except for the cattle farm located on the south-westward fringes that eats into the community land, no major conflicts were dictating everyday political life during the research period. Owned by a San Borjan citizen who does not reside there, the farm's premises are used to house an alternating number of cattle during the dry season every year (Figs. 4.1 and 4.2).[2]

Campo Bello's norms of resource use are regulated by two frameworks; the traditional norm system common to all Tsimane' communities within their official territory, and the non-traditional system imposed by the BBR mandate. As concerns the former, the land, the river, the shore, the forest and all game and plant resources within the village compound are under communal ownership and, as a common strategy practiced by many traditional societies, transformation from communal to private property can only take place through the input of human agricultural labour (see Godelier 1986). With regard to land, traditional communal norms in Campo

[1]School lessons held in the Tsimane' language were first introduced in 1987 and replaced by a bilingual school 3 years later. While the school teacher had been trained by the missionaries, the wooden school building was provided communally by the villagers. In 1993, after the issue of indigenous rights had become a crucial political concern in the region, the local government constructed a school building out of durable materials which is still operating today. The electric installations ceased functioning a few years ago and have not been fixed since.

[2]For more than a decade, the farm has been a constant source of conflict between the Tsimane' residents and the farm owner. According to the owner, the land was purchased before the Tsimane' were granted land titles. Legal procedures had been taken by both parties and, at the time of my second field research in 2006, were still ongoing.

Fig. 4.1 Bird's eye view of Campo Bello, 2004

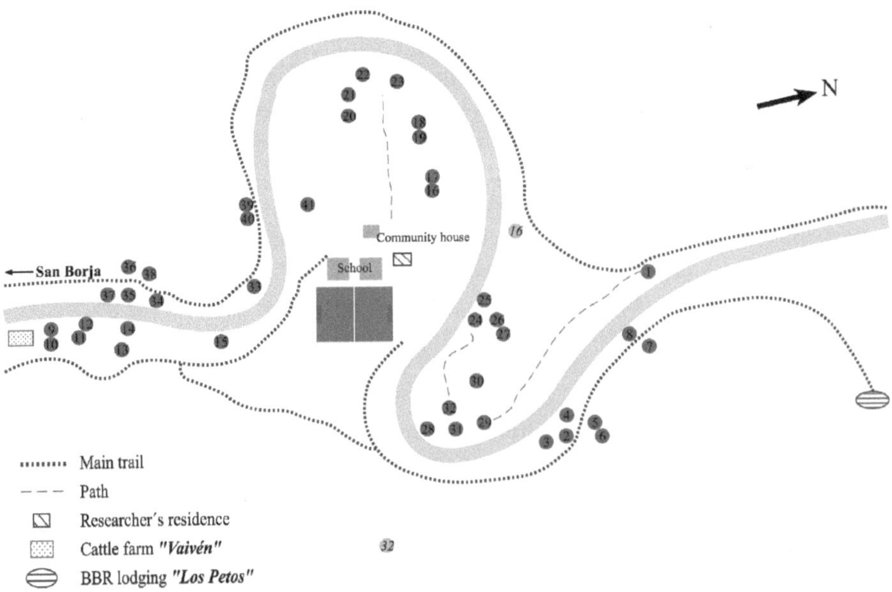

Fig. 4.2 Village map of Campo Bello, 2004

Bello impose no restrictions over forest clearance for gardening. Yet once a spot is decided on by community residents, it becomes theirs and people are automatically granted traditional tenure rights. Both, residents and passers-by from neighbouring communities have a meticulous memory about the ownership of cultivated and fallow gardening sites. Even after the final remains of an abandoned house have perished, fruit trees remain evidence of the earlier efforts of their owners. At times, they return to abandoned garden sites to exploit their harvest. I repeatedly witnessed how people became considerably irritated upon realising that their crops had been taken by others. Whereas all gardening sites are under private ownership, the school field, however, officially belongs to the community. Nonetheless, all rice and other smaller crop yields from this area are handed over to the teacher and his family. This gesture is not only a token of gratitude for the well respected teacher, but also an established village norm that was initiated with the arrival of the first teacher to the community in the late 1980s. Campo Bello's location within the buffer zone of a biosphere reserve requires the community to observe the rules and regulations imposed by the BBR mandate: to ensure an ecologically sound use of resources. For Campo Bello, these regulations do not contradict traditional norms and monitoring activities on behalf of the park rangers are largely restricted to the extraction of timber beyond subsistence needs.

4.2.1 Village Demography

What ought to be a fairly straightforward first activity when in the field, designing a demographic profile for Campo Bello, turned out to be quite challenging. Statements given by individual household members upon my initial rounds of household surveys rarely matched, and written documentation (e.g., a village census carried out by rangers from the biosphere park a year prior to my arrival) barely coincided with my own observations. What remained was a high degree of uncertainty. In retrospect, I believe that these complications arose essentially due to two aspects. First, families tend to leave out newborn babies when asked about the number of off-springs. Later I was told that newborn babies are hardly ever given names as they might not be strong enough to survive (Roman Durvano, pers. commun.). Only when a baby seems fit for life, does it become a fully integrated member of the family. Another difficulty I found was mobility. Though shorter trips are a recurrent feature in daily village life, some household members would embark on longer journeys, leaving the community for several days or weeks to engage in hunting, seize the occasional wage labour opportunity or simply visit kin upriver.

Looking back, the final demographic profile stems for the most part from personal observation and subsequent data triangulation. The latter entailed interviews with the teacher and the village head, as well as secondary data sources offered by the municipality of San Borja and the Great Tsimane' Council. Through this method I could solve the first constraint discussed above. The peripatetic nature of movement that characterises Tsimane' daily life did, however, still impose a methodological challenge for the purpose of the study, as some residents, albeit temporarily, ceased to be part of the community system of Campo Bello. Yet, I decided to include the total population in the biophysical study. This decision is based on the fact that the temporary out-migration observed was somewhat sporadic and took place for rather short periods. Likewise, it was not uncommon to see families upon their departure pack their canoes with nearly all their belongings including a variety of staple food stocks (plantains, rice, manioc and poultry). Rather than living off other peoples' food resources, people tend to return once their own food resources are depleted.

In 2004, Campo Bello had a total population of 231 residents, comprising 119 males and 112 females. The data reveals that age distribution resembles that of a fast growing society. What is striking is that we find a heavy bias towards infants and youths and relatively few old people. Moreover, general life expectancy is fairly low. In 2004, around 60% of the entire village population were below the age of 16%, 18% between the age of 16% and 30%, 17% between 31 and 60 years and merely 5% were beyond 60. If we now account for the fact that children start to fully engage in the work process around the age of 12, the informal workforce of Campo Bello amounts to at least 53%. This percentage rises if we also include the elderly population within our workforce, since they still participate in productive activities such as collecting firewood, fetching water and childcare, as well as providing general support for economic production. Shown in more detail in Chapter 5, children below the age of 12 also contribute quite substantially to the reproduction of the household economy.

The reduced numbers between the first stratum (0–5 years) and the second stratum (5–10 years) reflects a high infant mortality rate in the community. According to a baseline study carried out by the GTZ in the communities located within the biosphere reserve, average annual infant and adolescent mortality rates in 1997 accounted for 1.24% per Tsimane' family in the area, while mortality rates for Campo Bello were calculated at 0.58%. The gender ratio reveals a slight preponderance of males to females. An interesting observation is the fact that no women were found between 56 and 60, while women between 40 and 45 outnumbered their male counterparts by 1:5. Even though family members often spoke in a highly appreciative manner of the industriousness of their daughters, no actual gender preference was observed among the Tsimane'.

In order to study population growth patterns, various data sources were tapped into: the municipal census of 1994, a village census carried out by the village headman for the years 1997 and 2003, as well as primary research data collected for the years 2004 and 2006. Since only aggregate numbers exist for the former 2 years, reproductive histories could only be created for the latter 3 years. As indicated in Table 4.1, local population levels show an annual growth rate of 3.77%.[3] In the period I observed, roughly one third of this growth was due to in-migration and two-thirds due to births.[4]

[3]In comparison, the general growth rate of the Tsimane' population in the territory has been 4.86% during 1971–2002 (see Godoy et al., 2005b: 163).

[4]+7 could be observed for migration during 2003–2006 (immigration: 16, emigration: 9); +18 for reproduction (birth: 25, death: 7).

Table 4.1 Population growth, Campo Bello, 2006

Year	Household	People	Average/household
1994	23	107	4.7
1998	35	152	4.3
2003	39	210	5.4
2004	41	231	5.6
2006	42	235	5.6

Table 4.2 Population dynamics, Campo Bello, 2003–2006

	Total population 2003 210 (109 M/101 F)		
Immigration	Emigration	Reproduction	Death
14	3	12	2
	Total population 2004/5 231(119 M/112F)		
Immigration	Emigration	Reproduction	Death
2	6	13	5
	Total population 2006 235 (119 M/116 F)		

Despite their renowned 'taste for movement' (Ellis 1996), almost all Tsimane' families I met in 2004 still lived in the community upon my return to the field in 2006. During the period of my absence, the village population had increased by four, totalling then 235 residents. Table 4.2 reveals that four adolescent men had married and all resided with their wives at the male parents home.[5] One family had moved upriver to kindred and five residents, among them two recently born babies, had deceased during my 14-month absence.

4.2.2 *The Social and Political Sphere*

4.2.2.1 Village Hierarchy and Leadership

We have already learnt that the Tsimane' in any typology of comparative political systems are still a fairly egalitarian society, with the division of labour between men and women not engendering any idea of hierarchical disparity between the sexes. Albeit increasingly diminishing, in some communities shamans are still seen as traditional leaders and encountered with respect and awe. About 2 months into my stay in Campo Bello, I almost inadvertently found that Nativi was known to possess shaman-like healing powers and people would seek his services for cures from

[5]As two girls are local to Campo Bello, oscillating movements between the two sets of kin are very likely for the newly married couples.

minor ailments. While surely appreciated by his fellow community residents, some complained about his scarce community presence and general disposition. When I returned in 2006, he had moved upriver altogether. To my mind, his power status within the community diminished with the introduction of a formal political leadership system as a result of the village's official grant of community status in 1995. Formal political leadership is now in the hands of the village headman Roman, who was elected by general assembly and has held the post since 2003. Next in the village political structure are the mayor and the president of the school union. In quantitative terms, engagement of these official leaders in political affairs did not take up a lot of their time and effort at the time of my research. To my mind, one of the limiting factors may have been their fairly restricted command of the Spanish language.

Yet leadership attributes are not only restricted to formal status. To my mind, the teacher of Campo Bello, though being an ethnic outsider as he belongs to the Mosetene tribe, was generally respected as a local leader who people, including the village headman, would seek for advice. Known by everyone simply as the *profe* (teacher), his leadership skills became manifest in various ways. Beyond his role in holding together the school community, he was quite popular for his 'modern' ideas[6] as he encouraged the villagers to grow food for the market, he listened to complaints, tried to solve disputes and repeatedly hired men to clear his plots in exchange for money. He would provide for basic school materials for some of the students and tended to receive labour, plantains, game meat and fish in exchange from the villagers. The yields of the school plot, which the students clear, cultivate and harvest every year, are entirely handed over to the *profe* and his family.

As illustrated elsewhere in this book, individuals may also differ in the innate respect they evoke. Some individuals are considered stronger and more efficient than others, thereby enjoying a higher social status among community members. Ansermo, for example, was often talked about for his outstanding hunting abilities, while Juan was much admired for his agricultural skills. I also found that an individual's high standing is not gender-specific; likewise some of the local women enjoyed a particularly good reputation for making large quantities of strongly fermented manioc beer. The ability of these women frequently made their households the social centre of the village community, hosting beer feasts for thirsty kin people and neighbours. A high status is also attributed to entire families. At the time of research, I recall the Tayo family to be particularly reputed for their large-scale plantain production. My strongest memory of the Tayo home, in fact, is that of various kinds of green and mature plantain heads dangling from the rafters of their house. When there is a general lull in plantain harvesting, people pay frequent visits to the Tayo household in search of mature plantain cultivars. Whereas industriousness

[6] Despite his efforts though, he frequently whined about the Tsimane' limited sense of community: his modern ideas of moving the villagers closer to the school building in order to instil a somewhat communal spirit among the individual families were often met with incomprehension. When I returned to Campo Bello in April 2008, most residential dwellings had indeed moved closer to a newly built school building. Through his higher status within the community people trusted his ideas and followed his example.

and diligence are both desirable attributes that constitute a good spouse, laziness, on the other hand, is scorned upon unanimously. When Florencia was saddled with a lazy husband, for instance, her family and neighbours encouraged her to divorce him. It appears that if someone, husband or wife, is publicly recognised as a lazy person, it is absolutely tolerable to abandon him or her as he or she does not accomplish the productive role that is needed for the family's subsistence. Finally, Lorenzo's long absence from his family for wage labour elsewhere was looked at quite critically by others. Time and again, I heard people almost nostalgically comment on his former glory days as a village headman, when he showed concern and dedication to his people. What these examples tell us quite clearly is that the levels of respect individuals enjoy remain far from static; they require certain sets of behaviour patterns to be followed and thus undergo constant reproduction.

4.2.2.2 Communal Gatherings and Group Settings

While most of their working days are spent in solitary work tasks, the Tsimane' sense of community is undeniably instilled through cheerful communal feasts. The people of Campo Bello clearly enjoy such recreational gatherings and prepare for them with enthusiasm. The main annual village feast in Campo Bello takes place on the 4th of April every year, the community's official anniversary. Around this date, one can feel a certain festive mood hanging in the air. Even people from neighbouring communities drop their work to flock to the feast, where communal food is enjoyed, manioc beer and purchased alcohol flow in abundance, people dance and youths participate in sporting competitions. In recent years, the people of Campo Bello have gradually been adopting the celebration of national festivities such as Mother's Day, Father's Day and other national holidays. The occasional birthday celebration, an event mostly unheard of about a decade ago, has also gradually found its way into the community's repertoire of communal feasting. The celebration of a Wednesday and Sunday Mass service introduced by the New Tribes missionaries represents another communal event in Campo Bello.

Just as in other Tsimane' communities, communal work constitutes another group setting in Campo Bello. Activities involve the cleaning and general maintenance of the school building and the clearing of communal paths. In 2005, the construction of a communal sanitary post, as part of an area-wide development project, has also required communal effort. Despite the toil of the tasks involved, group work appears to have social rewards. This was brought home to me on one particular occasion when, on my daily rounds through the forest, I came across a small group of local men engaging in communal path clearing. They seemed to be enjoying themselves, with jokes flying back and forth, provoking jovial laughter among the men. Santos, the eldest member among the group, asked me, courteously but determined, to join his wife in preparing a common meal that would be awaiting them upon finishing their hard work. I believe it was a polite but clear indication of the men wanting to be among themselves, and, within the frame of playful sociability, enjoy this short period of male group interaction.

4.2.3 The Village Economy

Campo Bello is characterised by both, a monetary economic system (marketing of agricultural crops, forest products and wage labour) and by reciprocal kinship relationships of barter.[7] Similar to other communities in the area, Campo Bello's economy rests upon four main pillars. While agriculture takes up the central place in terms of time use, labour input and monetary output, fishing, foraging, raising domestic animals and occasional wage labour opportunities

Fig. 4.3 Successful hunt of a wild boar, 2004

---

[7]In line with Polanyi's (1944) argument, in traditional societies the village economy does 'not exist as a separate entity but is rather embedded in other institutions such as kinship or religion' (in Godelier 1986: 181). Indeed, what he terms as the dominant principle of the integration of the economic process, namely reciprocity, is also an important aspect of the village production system.

Fig. 4.4 Fishing in the Rio Maniqui, 2004

complete the local economic profile. In line with Netting's (1993) arguments, the Tsimane' represent a typical smallholder society that combines one dominant economic strategy with various other more subordinated survival strategies. Slash-and-burn agriculture is practised by all households in the village and presents the main source of monetary income for most.[8] Rice, as the main cash crop, makes up for approximately 40% of all cash income in the community. The cultivation of rice is also the most time and labour-intensive agricultural task in which all able family members are involved at some stage. Apart from rice, the villagers also plant plantains, maize, manioc and other crops of smaller importance, such as peanuts, sugar cane, citrus and varieties of sweet potatoes.

Besides agriculture, fishing takes a central place in village life and comprises the main source of protein in the peoples' diet (Fig. 4.4). At various seasons of the year, entire families engage in fishing trips, either to the river, the nearby lagoon or to one of the smaller streams flowing into the Rio Maniqui. Local families use a variety of fishing techniques ranging from hooks, bow and arrows to more elaborate fishing nets. While communal *barbasco* fishing is still an activity observed

[8]Average annual income per capita amounted to $295 USD in 2004. The minimum annual income per capita in Bolivia amounts to $980 USD (Vadez et al. 2005).

Fig. 4.5 Gathering forest foods, 2004

in more remote Tsimane' settlements, I have not observed any events of this sort in Campo Bello. This is probably due to recent efforts on behalf of the BBR staff to completely abandon this kind of fishing technique as it may carry the health risk of water contamination. Livestock raising is only of minor importance to the community. While all families own at least some poultry, only one family also owns cattle. Hunting and gathering activities complement the local economic profile and are practised to varying degrees by individual community members (Figs. 4.3 and 4.5). As a general observation, many of the older men were enthusiastic hunters while a growing number of younger men were more enticed by erratic wage labour opportunities outside the community. At the time of research in 2004, only Lorenzo had left the village for 2 months to work for logging companies upriver. Of all wage labour activities male residents engaged in, about 40% worked as agricultural farmhands, 15% for cattle ranchers, 4% for logging firms and 12% were involved in other activities.

4.3 The Sociometabolic Profile of Campo Bello

This short overview of the general community setting has intended to pave the way for the following metabolic study that captures the essence of the book: the description of the metabolic relationship between the people of Campo Bello and their natural environment. How efficient is the peoples' current use of material, energy and land and what are possible future pathways? To this end, I am going to apply the concepts of social metabolism and colonization both of which have their own accounting tools. As concerns the former, the tools applied in this study are Material Flow Accounting (MFA) and Energy Flow Accounting (EFA) in order to obtain a full balance of the material and energy exchange relations between the local community and its environment. With respect to colonization, I am going to calculate HANPP, an indicator that measures the impact of peoples' land use practices. In light of this, the remainder of this chapter starts by presenting the data generation process in the field, followed by a detailed and systematic presentation of the metabolic results. I first examine the biophysical stocks and subsequently move the focus to the material inflow and outflow categories. Following the same logic, I then present the energy flow results. Once this individual examination comes to a close, the pieces are put together into a comprehensive material and energy flow diagram which, through aggregate indicators, reveals the local pressures on the environment. Only then the focal point shifts to the impacts of land use.

4.3.1 Field Methodology on Stocks, Flows and Land Use

Equipped with questionnaires, a measuring tape, a compass and other items a young and ambitious student considers essential for field work, I set out to do the field investigations. Following the systemic logic of the metabolic concept, I first of all specified the material stocks of Campo Bello in order to provide a clue on the 'weight' of the social system. The stock account comprised the local human and livestock population, the housing infrastructure and all those artefacts that have been part of the system for at least 1 year. Getting started meant drawing up a baseline for the compilation of socio-demographic information. This data included name, age, date and place of birth, years of residency in the settlement and self-perceived ethnicity. Obtaining accurate data sets on age, however, became quite a complex endeavour, since some of the older community members simply lack written birth records. To overcome this problem, I tried to piece together information from various sources such as school records and comparisons with other family members. I then established an inventory of the local livestock population and all man-made structures (e.g., houses, wells and latrines). What had to be done next was to calculate the respective weight of the stocks in metric tons. To calculate the mass weight of the local population, a small representative sample of residents was weighed in order to subsequently calculate the average body weight and extrapolate it onto the entire village population. Weight estimations for the livestock population were derived from assorted available written sources (e.g., CIMAR 2001).

As a general observation, calculating the weight for some infrastructures and artefacts meant a challenge that in fact required creative solutions. But I was not alone in this dilemma; two of my colleagues, Clemens Grünbühel and Simron Jit Singh, had been confronted with quite similar field challenges in the years before. I thus gratefully followed their advice to use, whenever possible, a categorisation by size (e.g., small, medium and large) before appointing an average weight to each of these types. Attributing weight to housing structures, for example, entailed a calculation of the material ratio by measuring each of the raw materials used. To make this process technically feasible, I established the following categorisation of five different types of houses: type OT (old traditional), type NT (new traditional), type NT+D (new traditional plus door), type NT+DM (new traditional plus door and mud) and type M_{odern} (traditional and non-traditional material mix). The housing stock account was carried out during my main research period in 2004 and subsequently repeated in 2006 in an attempt to examine the stock changes that had occurred since the previous undertaking. Concerning household artefacts, only a minor fraction of household utensils or agricultural tools could be weighed physically. As for all bulkier artefacts, however, I had to resort to secondary literature sources or personal estimates.

Following the calculation of the system's stock account, my second consideration was the definition of the territory the local community is entitled to exploit. This is what we refer to as 'domestic environment' (forest areas, agricultural land, river and savannah areas). Bearing in mind the two parallel boundary distinctions (social/natural, social/social) I started to collect data on Campo Bello's material flows. On the inflow side, these entailed the materials extracted from the domestic environment (social/natural) and imports from other social systems (social/social). On the outflow side, material outputs discarded into the domestic environment were equally accounted for as exports to other social systems. The generation of material flow data took place during various stages of my research period and entailed a methodological mix of interviews, questionnaires, measuring, weighing, and participant observation. The measuring unit used for material flows was tonnes/year and all biomass flows were calculated either in terms of weight of dry matter or fresh weight when harvested.[9]

As to inflows, domestic extraction from agriculture made up the bulk of extractive activities and was calculated relying on a variety of sources. In a first step, data was acquired by interviewing producers who were asked to provide data on recent harvests. Informants indicated how many bags of rice, manioc and maize they had harvested in 2004. The answers were given in the locally common measuring unit

[9]Except for grass and timber when removed from woodland, all other materials were integrated in the flow balance with their fresh weight when harvested or traded from another system (see Matthews et al. 2000).

arroba (one *arroba* equals 11.4 kg). As for plantains, which are harvested once they are mature, Tsimane' informants gave estimates of the approximate cultivars harvested every month. People usually give accurate accounts on rice harvests as rice is a cash crop that is generally harvested in one go. Concerning manioc and maize harvests, estimations were presumably less accurate as both crops are commonly harvested more frequently, in smaller quantities. Following the interviews, all active agricultural plots were measured and account was taken of the number of plants per 10 m^2. In so doing, it was fairly easy to arrive at the area productivity of the crops and compare yields with producer statements. The final account concerned fruits and vegetables both of which are harvested at all times throughout the year for home consumption only. Hence to obtain proper yields, the number of trees and the area of vegetable gardens under production were used as a base for calculating production figures (see Singh 2003; Mayrhofer-Grünbühel 2004). Also, figures derived from the food sampling method were cross-checked in order to arrive at precise consumption figures.

In order to obtain data on the food consumption of households (including the consumption of fuel wood, water and animal fodder), I opted for the food sampling method. To this end, five local households were selected to conduct an in-depth study on their food production and consumption behaviour on three consecutive days.[10] On the day designated, I turned up and placed myself in a corner from where I could observe the food preparation and consumption process. Individual household data was subsequently analysed and extrapolated onto the entire village population to arrive at average per capita food intakes. During the sampling, all raw food and firewood to be used were weighed for a total of 15 observation days and then divided by the number of family members (sex, age) sharing the meal.[11] After that, the entire food preparation and cooking process was monitored, including the actual disposal of left-over food. Again, Singh's (2003: 113) advice, to use a food list containing domestically extracted and imported foodstuffs, served as a welcomed aid. The food sampling covered the dry and wet seasons in order to capture variations in seasonal resource availability. The fact that hunting, gathering and fishing yields are almost entirely consumed locally, the food sampling method provided a suitable tool for deriving consumption figures. Since hardly any households frequently feed their livestock, I had to rely on secondary data on animal food consumption which was either derived from literature (CIMAR 2001) or through personal communication (Fridolin Krausmann 2006). Finally for estimating the use of construction wood, data was obtained from household interviews as well as personal observation of house building, repair work or general maintenance activities.

[10]According to Grünbühel (in Schandl et al. [2002: 81]) 'households should be carefully selected and should present the entire social continuum of the considered community'. My choice of households was more erratic and rather determined by which households were prepared to offer me an insight into their daily consumption patterns.

[11]The number of consumers per household was derived by counting all adults of both sexes as whole consumers, while all children under 10 were counted as half-consumers.

The second category concerns the calculation of imports from other social units. For this purpose, I made a distinction between biomass (several food items, livestock), fossil fuels (kerosene), household artefacts (durables and consumables) and minerals (salt). As there was no single entry point for imported goods (i.e., shops), data was gathered during a quantitative household survey covering all households, subsequently cross-checked with interviewing itinerant traders at different times of the year and supplemented by a great deal of personal observation. The category 'durables' could be calculated quite precisely, since a second artefact stock account was established shortly before leaving the village in February 2005. The newly generated data could thus be compared with the data set gathered at the onset of the field study several months earlier. In a similar endeavour, a third stock account was drawn up upon my return to Campo Bello about a year later and was weighed against earlier data sets.

On the outflow side, wastes and emissions excreted into the environment were calculated using a variety of data sources. While bio-metabolic wastes were derived from technical literature sources (Spedding 1975; Pfister 2003), other wastes and disposals onto the natural environment were obtained through direct observation and household interviews. Material exports to other social units were derived using the same approach as with traded imports. Data was thus generated through quantitative household surveys and subsequently cross-checked with interviewing itinerant traders. A great deal of personal observation was just as useful for deriving grounded estimates. In order to reckon the category 'deliberate disposals', two *arroba* (22.8 kg) were calculated for the cultivation of one ha of rice and maize. For manioc, one *arroba* was calculated as an input into the natural system for the cultivation of the same area size, while the production of plantains requires the input of 1,000 plantlets for the same area. Unintended post-harvest losses were calculated assuming a loss rate of 15% of the entire agricultural harvest.[12] Finally, the UNFCCC (1998) handbook provided the source for calculating emissions.

Once the data for calculating material flows has been generated, the calculation of energy flows usually becomes a much lighter undertaking. This has to do with the fact that the conversion of biomass and fossil fuels from mass weights to energy units is largely based on secondary literature research. For the purpose of generating comparable data sets with other local case studies on energy flows, the energy units used were calorific values. Most values were derived either from specific literature (FAO 2001; Schandl et al. 2002; Pfister 2003) or personal communication (Krausmann, Schandl, Singh). The conversion of edible weight for each food type into energy units is based on the food consumption table compiled by the Institute of Nutrition of Central America and Panama (Wu Leung and Flores 1961). For estimating the amount of useful energy delivered, I followed the

[12] This figure has been calculated from information gathered during interviews as well as personal observation. Households often lack suitable material for adequate storage of field crops and people often complain about their inability to protect their produce against rats, insects and other bugs.

estimates suggested by Haberl (2002: 74) and Singh (2003: 115). To calculate the efficiency of fossil fuel (kerosene) in producing light, a factor of 1% was applied, while the process energy from fuel wood was taken to be 25%. Likewise to determine human labour efficiency, a low estimate of 50 W was taken and multiplied by the amount of working time per capita per year.[13] Due to the rather insignificant values on power expenditure, no data was collected on the consumption of the solar energy plant that was set up in the community by the national telephone company in 2001.

My next endeavour was the examination of land use. For the purpose of a metabolic analysis, what is crucial to take into consideration here is the fact that territorial boundaries are culturally defined and demarcation lines, therefore, are not automatically congruent with the territory's geographical or administrative boundaries. Admittedly, the definition of the territorial borderlines of Campo Bello was no easy undertaking and caused a certain level of uncertainty, at least at the beginning stages of research. This is mainly due to the fact that the official land titling process was still ongoing at the time of research and the village headman had revised his estimates of the community's border upon my second visit in 2006. Nonetheless, with the support of oral statements by local people and the fact that demarcation posts actually existed at some strategic places, I was able to delineate the village boundaries upon return in 2006. Accompanied by a small group of locals and GPS technology, the distance between these demarcation posts was hence measured. Once the territory the people of Campo Bello are entitled to exploit was defined, the next step was to account for the different ecosystem and land use areas in the community (primary forest, secondary forest, agricultural area, water body & beach, dry savannah, humid savannah), followed by the types of total land cover (housing area). Data was derived either from official statistical sources (land cover images provided by the departmental research institute CIDDEBENI and municipal development plans) or calculated manually with a measuring tape (in case of agricultural and housing areas). To obtain accurate results on the latter, I also used questionnaires combined with qualitative interviews with household heads. The questionnaires were designed to give information on the species cultivated in kitchen gardens, the number and size of active fields, fallows and the cropping cycle. Measuring the family's surrounding kitchen gardens and agricultural fields took place in parallel. Again, the physical measuring of these areas was a communal undertaking which could not have been realised without the valuable help of my Tsimane' hosts.

In order to calculate the intensity of the colonization strategies in Campo Bello, HANPP was applied. To generate data on HANPP is a five stage process and was done the following way: to first of all obtain a rough estimate, the NPP of potential vegetation (NPP_0) was calculated by using the Miami model (despite its reported flaws, see Haberl 2002), based on average annual temperature and precipitation

[13]According to Smil (1992) in Haberl 2002: 74), humans can deliver up to 100 W in continuous work. As this is, however, a maximum, losses have to be taken into account.

rates (see Lieth and Whittaker 1975 in Haberl 1995). For more precision, I estimated the different land cover types in the absence of human intervention. To do so, it was assumed that, in the absence of human intervention, the actual patches of secondary forest would have remained primary. Other naturally occurring land cover types (i.e., savannah areas and water body) were estimated with roughly the same NPP values as those taken for the calculation of actual land use. I based my latter assumption on the fact that the local site had been populated for less than a century and people use extensive land practices only. Hence, many parts of the local system were still fairly virgin. An aerial land use map dating from the early 1970s supported my assumptions and aided the calculation process. The actual NPP (NPP_{act}) was derived from aerial land use and land cover images of the area that were obtained from the regional research institute CIDDEBENI. Data on annual biomass harvests had already been calculated for the material flow account. The indicator HANPP was then derived by subtracting the NPP_h (i.e., harvested NPP) from NPP_0. To increase the level of comparability with other local cases, NPP was calculated in energetic terms (TJ). Finally, it should not go unmentioned that due to the general limitations of reliable data sources, my HANPP analysis entails the calculation of above-ground NPP only.

To my mind, this section would feel incomplete without some personal reflections that emerged at various stages of the data collection process. As a general observation, applying the range of biophysical methods in the field was quite am enjoyable endeavour. I remember my constant measuring and seemingly unending questions like ‘How many logs of firewood to you gather in the forest?’ Or estimating the weight of piles of clothing, which was to many a source of amusement. Looking back, I believe that being the only outside researcher in Campo Bello helped a great deal in building a rapport with the local population. My efforts to pick up the local language not only became a reliable source of laughter, but also substantially aided an understanding of the local context.

At the same time though I became well aware of the drawbacks of some of the field methods applied. First and foremost, the food sample method has, by some been condemned as ‘an intrusive observational method’ (Byron 2003: 54). Admittedly, the researcher’s presence during the meals undeniably introduces bias. Adding to this, weighing every gram of food during 30 meal samples is equally peculiar and, while technically feasible, is in my view not an alternative for reasons of simple decency. I was thus left with the uncertainty of how to handle this process in order to arrive at scientifically valuable food data. What I opted for in the end was a less intrusive food sampling method. First, food samples were only physically weighed in the first household; as to the remaining households, weight estimates were resorted to. At the same time and in line with Bernard’s (2006: 354) advice, ‘Presence builds trust. Trust lowers reactivity. Lower reactivity means higher validity of data’, nutritional data collection was only carried out after months of trust-building. All households where food sampling took place were indeed befriended households with whom I had established a good rapport in the months prior to this activity. One of the major trade-offs with food sampling is the fact that some foodstuffs go unreported. This is the case with fruits, for example, that are eaten away from home on various occasions throughout the day. An attempt to overcome these methodological difficulties was made by conducting additional interviews with household heads coupled with a great deal of personal observation during my recurring daily rounds.

Another reflection concerns the flaws of recall questions that were part of the household surveys. Here, the main measurement errors may have occurred when people reported on the sale of agricultural crops and corresponding cash incomes. People may simply forget things that happened in the past or biases occur when research subjects intentionally over- or underestimate certain variables (see Reyes-García 2001; Bernard 2006). Prior to field measurements for example, Tsimane' informants were asked to estimate field sizes. More than 62% of all respondents, however, underestimated their own field sizes. To overcome errors and biases I followed the advice given by Reyes-García (2001: 20), namely to insure the presence of other family members during the interview situation. This was an opportunity for all to join the event, support the interviewee and, if necessary, correct the information given. As research progressed, I became more attentive to peoples' periods of absence from the village, the arrival of river traders and other important events. Admittedly, the fact that my first translator was the village head proved extremely useful, as he was informed of all happenings in the community. He provided me with necessary details that would otherwise have been omitted from my research.

Gender aspects equally deserve attention. At the outset of my research, the local women were rather shy and upon my visits, when only female household members were at home, I was frequently told to return at a later time when their husbands would be present. My research interest, however, also concerned women (e.g., trading or housekeeping activities). Only with the passing of time, most women grew more confident with me and gradually opened up. In a similar fashion, being a female researcher may not only have limited my access to certain information, but also influenced my perception about others (see Scheper-Hughes 1983). For example, when asked about monthly imports, men hardly ever included the purchase or trading of alcohol. Some might have found it inappropriate to discuss this issue with a woman. As a final reflection on gender, my research design may have introduced, at least at the beginning stages of research, a certain level of male bias. When collecting data on material assets, for instance, male respondents provided most of the information.[14] I did not bear in mind, however, that some items (e.g., poultry) are entirely owned by other household members who would, indeed, have been more suitable interviewees. The same goes for the marketing of forest products that

[14] For the underestimation of working time, Pastore et al. (1999: 341–343) raises the problem of reliability of self-reported assessments. He claims that work invested by children and the elderly does not get the same consideration as the time invested by adults. Also, in households were the opportunity cost is low, the distinction between working and non-working time becomes somewhat blurred.

is to a large extent carried out by women. It was only after acquiring a more profound understanding of gendered time and labour allocation, that male bias was largely eliminated.

Finally, the fact that I was an individual researcher in the field may have had an impact on the data quality. In classical anthropological literature, true fieldwork is unanimous with a 'solitary endeavour' as, in Malinowski's (in Werthmann 2004) view, it demands complete immersion in the world of the studies group. Without questioning his assertion, I still believe that the use of teams would have made the research process more effective, at least at some stages. During the initial exploration phase, for instance, a team approach would have most likely provided a framework for comparison and theorizing. Second, a research team would have facilitated much bigger research samples in order to provide a higher level of data validity. Nonetheless, I did enjoy the sporadic company of other researchers who, based in San Borja, engaged in different but still somewhat related research with the Tsimane'. Our frequent exchange of information and mutual interest not only helped me to expand my own horizon and interpretation of things, but surely also helped to put my own findings in a broader perspective and identify mistaken beliefs (Fig. 4.6).

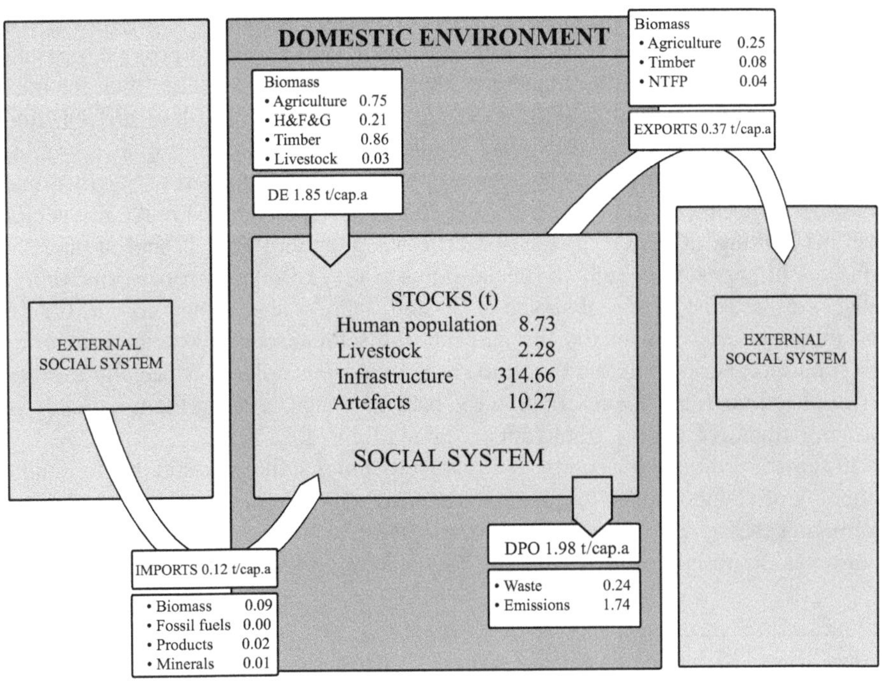

Fig. 4.6 Material flows in Campo Bello, 2004

4.3.2 Local Material Stocks

4.3.2.1 Human and Livestock Population

In 2004, the mass weight of the human population was calculated at 8.7 t. For livestock, a total of 480 chickens, 18 ducks, three pigs, five cattle,[15] 54 dogs and 39 cats were counted, accounting for a mass weight of 2.3 t. While all families raise a few chickens, fowls and ducks – the average numbers per household being 12.1 poultry – at the time of research only one household possessed three pigs, while another family owned five cows. I was told by an elderly informant that not so long ago, various local families used to own pigs, yet most became increasingly angered by their incessant digging out of root cultigens. People became particularly irritated when manioc crops, such a valuable ingredient for home-made beer, were uprooted. During the daytime, animals are allowed to forage freely for insects and seeds in the environs of the household and are sometimes fed rice husks or maize in season.

Raising poultry is very popular among Tsimane' families as it involves minimal labour investments and provides for additional meat stock when game meat is scarce. Also, poultry is particularly easy to transport and sell at the market in San Borja. Cattle breeding is an extremely rare activity among the Tsimane' inhabiting the lower Rio Maniqui banks. In 2004, the five cattle grazing outside Enrique's home in Campo Bello were a remnant of a development effort implemented by the Tsimane' Council some time back. It consisted of temporarily 'renting out' a bull and a cow to those Tsimane' families in the territory who would provide a certain area of pasture. These cattle though are kept as a saving asset rather than for meat production. Dogs are not raised for eating but valued for their hunting abilities. I was also told by Lorenzo that every hunter possesses his own dog as a hunting companion and people (not even close relatives) are not supposed to borrow hunting dogs from others. Breaking this rule, so he recounted, would make the person lose his hunting skills. Of all the domesticated animals found in and around the house, dogs appeared to enjoy a special status as they receive a proper name that in most cases, at least as was observed in Campo Bello, refers to their colour. Parrots and monkeys are occasionally brought into the settlement and kept as pets, often just for a short period before they escape back into the woodland. Especially children affectionately care for them, clipping the parrots' wings, feeding them patiently or leashing monkeys so they would appear tame (Table 4.3).

In summary, animal husbandry is a relatively minor activity within the household economy of Campo Bello. In quantitative terms, this also becomes manifest when examining the weight ratio between people and livestock, with the former almost four times the weight of the latter. The Tsimane' do not make use of animal labour

[15] There are also 80 cattle on the farm owned by a San Borjan citizen, which is located within the limits of the community. These cattle only grazed in Campo Bello during the dry season in 2004 (May–October) as for the rest of the year they were kept in a different community. These cattle, however, are not part of the social system and have hence not been included in the stock account.

Table 4.3 Human and livestock population, Campo Bello, 2004

		2004
		(t)
Human population		8.72
Subtotal		8.72
Livestock		
	Cattle	0.79
	Pigs	0.18
	Poultry	0.79
	Others (dogs, cats and pets)	0.52
Subtotal		2.28
Total		11.0

for economic production except for dogs who are an indispensable asset for hunting trips. As concerns the reasons for keeping livestock, motives are threefold: firstly, farm animals are mostly kept as a saving asset and poultry is conveniently transported to San Borja, where it is sold or exchanged at the market. Meat production is the second, but arguably less important motive. While the consumption of farm animals was highly solicited at village feasts and people would freely donate a chicken or two for the communal soup pot, I sensed a rather different situation within the intimate confines of the household. There, women frequently voiced their hunger for forest meat, urging their husbands to set out on a hunt in order to satisfy their spouses' demands. Only on occasions, and much to the disappointment of the women, when no forest meat is available, families resort to the consumption of poultry or duck meat. The third motive for keeping livestock is simply for pleasure as some wild animals were kept as pets.

4.3.2.2 Housing Infrastructure

Infrastructure comprises the second category of Campo Bello's biophysical structures and primarily consists of communal and individual housing structures. In 2004, the infrastructure in the village constituted a total of 49 residential huts, 27 kitchens, 16 hen houses, one school building with an additional teacher's home attached to it, one school toilet, 21 individual steel latrines and seven concrete wells.[16] Communal buildings were limited to the concrete school premises (including the additional teacher's building) and the traditional community house located just a stone's throw away from the school premises. While the concrete school was built by the municipality of San Borja in the early 1990s, the community house is a traditional construction erected by communal effort. It shelters meetings of all sorts and is the place where

[16] Two families have dual residences; one is the BBR; the remainder are abandoned. The additional building on the school premises was originally constructed to host the teacher's family. The family, however, did not inhabit these premises but constructed their own home nearby.

village feasts take place. The wells (mostly shared by extended families) and steel latrines are remnants of an international development project implemented in 2001.

The most common type of the Tsimane' house is a sturdy, rectangular structure, about 6 m long and 4 m wide, with a palm-leaf roof that peaks at about 4 m off the ground. It often consists of uva grass[17] walls and a single doorway. Most homesteads have a separate kitchen structure, often a bit smaller than the house, and usually one, occasionally two hearths, depending on the number of women with children living in the house. Walking outside the house, one finds a patio that extends in a rough circle about 10–15 m around the dwelling. The clearing is swept daily and carefully kept free of weeds; a preventative measure against poisonous snakes that approach the housing area especially at night time. The patio area is part of the house, for it serves as a yard in which various domestic chores can take place throughout the daily round.

Erecting a Tsimane' home is no small undertaking and demands both heavy labour input and strict attention to detail. The construction of a medium-sized dwelling requires some 2–3 weeks, including the gathering and assembling of the materials used from the forest. The duration of the construction depends not so much on the dimensions wanted, but more on the number of men collaborating. The situation I mostly observed was the collaboration of two to three men, in many cases the head of the household accompanied by one son or son-in-law. Women also partake in the house building process, though their tasks are mostly limited to fetching and weaving palm fronds for roofing or obtaining the balsa wood (*Secropia* sp.). The construction process starts with four to six corner posts of dense hardwood (*Astrocariyum* sp.) being set in ditched holes. The men then carve notches in the top of each post to cradle the main beams that become the two long sides of the base on which the roof will in due course be placed. Subsequently, the cross-beams are laid across the main beams. A traditional Tsimane' house is constructed without nails; instead, the bark of balsa wood is used. Lashing is the main method by which poles and other components are held together. In fact, much of the regular maintenance work required on older houses is the repair of worn out lashes.

For roofing, the Tsimane' use two varieties of palm leaves: *jatata* (*Geonoma diversa*) and *motacú* (*Sheelea princeps*) which are lashed to the rafters to complete the house. *Jatata* roofs are more resistant to both rot and parasites and the way they are fastened ensures that the roof becomes completely watertight. Such a roof provides for durability and may last up to 7 or 8 years, in fact sometimes longer than the house posts which may begin to decay at the base after 5 or 6 years. The posts though, still last a few more years before definitely compromising the structure's stability. Notwithstanding, *jatata* only grows along the upper Rio Maniqui and is therefore rarely obtained by the villagers of Campo Bello. In 2004, apart from the school building only five individual households had used *jatata* panels for roofing, all of who travel regularly upriver to visit kindred. The palm variety most commonly

[17] Uva grass is a sturdy but light-weight bamboo-like cane that grows wild in stands along the river banks.

used for roofing in Campo Bello is *motacú*. While found in abundance all year round, this palm species is less resistant to disintegration. The general lifetime of a *motacú* roof hardly ever exceeds 3 or 4 years, and the occasional repairs do not extend its life by much. The steady smoke generated from smouldering logs (especially in kitchen structures) helps to protect the roof from plant-eating bugs and other parasites. Roofing one house requires about 300 *motacú* leaves, 150 on each droop, which are braided together to prevent the roof from leaking. Time and again towards the end of the dry season, faulty dwellings are repaired and leaking roofs fixed in order to withstand the forthcoming heavy rains.

Except for the school premises, all residential huts have been constructed with traditional materials: wood, uva grass and palm leaves for roofing. In 2004, out of a total of 76 individual homes, 20 had no walls (type OT), while 27 kitchen structures also belonged to this housing type. Concerning the second category (type NT), 14 houses consisted of walls of uva grass poles. The walls usually cover all four sides and in the middle of one long wall the Tsimane' usually place an opening which serves as a door. With respect to the third category, 13 houses with a wooden door (*Swietenia macrophylla*) were encountered. Byron (2003), who observed locks on several Tsimane' dwellings in a neighbouring settlement closer to the market town of San Borja, argues their significance as a sign of acculturation, since it was never a common practice among Tsimane' families. This habit, however, had not yet reached Campo Bello during my research period. Finally, two mud houses with doors were found among the residences and subsequently categorised into type NT+DM. The application of mud is a labour intensive activity as only *jak pirij*, a certain clay-dominated type of sticky earth found along the river shore, is suitable for use. In terms of total land cover, the built up area and its related infrastructure accounted for only 1,864 m^2 in 2004. In 2006, imported material structures had increased slightly due to the construction of a concrete biosphere reserve lodge that was placed on the very edge of the community. The modern lodge was to substitute a former wooden construction where park rangers would be able to withdraw on their regular night patrols.

In his biophysical study of the small island of Trinket, Singh (2003: 118) asserts that interventions from government or other development actors have indeed caused changes in individual housing patterns and brought additional land under construction. This has not been the case for Campo Bello. From what I have gathered, I indeed believe that individual housing patterns are unlikely to change, at least in the near future. First, a look at local history reveals that placing a concrete school in the community in 1993 did not influence the way individual houses are built. Needless to say, imported bulks of construction material can hardly be afforded by any of the local families, while at the same time local building material exists in abundance around the community and can be freely accessed. It also seems that the people's renowned 'taste for movement' does not support sophisticated housing structures. In addition to these arguments, it should also be noted that modern corrugated iron sheets for roofing are not necessarily apt for the region as they produce immense heat and would make hot summer afternoons unbearable.

In terms of weight, Table 4.4 provides two indications. On the one hand, it illustrates the mass weight of traditional and non-traditional building materials in infrastructure

Table 4.4 Infrastructural stock increase, Campo Bello, 2004–2006

	Type of material	Total weight in 2004 (kg)	Total weight in 2006 (kg)	Material ratio (2006) (%)	Increase (%)	Material p.c. 2006 (kg)
Traditional						
	Wood	97,368	107,597	31	10.5	450
	Uva grass	6,240	7,318	2	17.2	30
	Palm leaves	72,700	80,800	23	11.1	340
	Domestic clay	5,280	5,280	2	0.0	20
	Subtotal	181,588	200,995	57	10.5	840
Non-traditional						
	Clay (bricks)	57,720	61,560	18	6.6	260
	Cement	11,304	13,128	4	16.1	60
	Gravel/sand	64,051	74,387	21	16.1	320
	Steel	735	750	0	2.0	0
	Subtotal	133,810	149,825	43	12.0	640
Total		315,398	350,820	100	11.3	1,480

and points toward the material growth of local stocks, on the other hand. In 2004, the mass weight calculated from the bulk of traditional building materials accounted for 181.5 t, while biophysical structures made of non-traditional construction material amounted to 133.8 t. What is striking is the heavy weight of bricks and sand particularly as compared to much ‘lighter’ traditional building materials. In terms of material growth, the total built infrastructure, which in 2004 had been 315.4 t, had increased to 350.8 t in 2006. This illustrates a total increase of 11.3%. When examining the weight ratio of traditional and non-traditional building materials between the years 2004 and 2006, we obtain fairly equal growth numbers, 57% and 43%, respectively. As an interesting footnote, it should be noted that the substantial material growth did not automatically entail noteworthy land cover changes; a mere 8% increase in the built up area was in fact observed.

In his study of an agricultural village in Laos, Mayrhofer-Grünbühel (2004) makes an interesting attempt to explore social differences through material consumption (e.g., through the type of housing) and concluded that there exists a strong correlation between the availability of resources and the quality of housing. I did not observe this phenomenon in Campo Bello as no relation was found between material wealth and housing patterns. On the whole it seemed that only minor importance was given to housing structures since a family’s status appeared to be more related to other status symbols such as bicycles and large radios. Still, housing patterns are slowly changing as younger generations tend to build houses with uva grass walls and wooden doors. Another observation rather curious to an outsider is the fact that Tsimane’ men often prefer to build new dwellings after 2–3 years instead of carrying out regular maintenance work. Sometimes, old dwellings are left to decay. Most often though, they are taken apart and posts are used for the construction of new homes.

4.3.2.3 Household Artefacts

To the newly arrived visitor, a Tsimane' home may seem rather empty (Table 4.5). Except for a couple of wooden benches and palm mats spread out on the dirt floor, what most strikes the visitor is the orderly collection of manufactured bows and arrows that are neatly stored on the rooftop just above one's head. On his travels through Tsimane' territory at the turn of the last century, the Swedish explorer Erland Nordenskiöld (2001 [1913]) lamented the difficulties he had in engaging in barter trade with the Tsimane' as, so he observed, families would possess a limited range of commodities only. Only when the looming need for a new hunting device or household utensil arose, were the Tsimane' prepared to allocate the time and energy needed for its manufacture. Still today, a Western observer may probably miss what in our world view is considered an 'object of decoration'. A second glimpse though, and this time with a more attentive eye to detail, reveals a different picture. Bones or skins from hunted animals, 'trophies' or souvenirs from past hunting expeditions, are mostly kept in bags hanging from the roof of their houses rather than openly displayed. I frequently observed babies with beautifully beaded amulets made of animal bones attached to their wrists. According to Ellis (1996), these trinkets are considered a lucky charm and contribute to the baby's future strength.

As befitting a traditional society, it is not striking to find that the majority of artefacts owned by the villagers were produced within the community (87%), while the remainder were imported. In 2004, canoes constituted the bulk weight of traditional assets and were owned by 21 households. The enormous increase in weight of wood (113% between 2004 and 2006) can partly be explained by the fact that people had to manufacture new canoes after many got destroyed during the river flood in early 2006. Concerning non-traditional materials, the largest increase in weight was that of steel, followed by plastic (23.3% and 19.1%, respectively between 2004 and 2006). The increase may again reflect the impacts of the flood, since food

Table 4.5 Artefact stock increase, Campo Bello, 2004–2006

	Type of material	Total weight in 2004 (kg)	Total weight in 2006 (kg)	Material ratio (2006) (%)	Increase (%)	Material p.c. 2006 (kg)
Traditional						
	Wood & uva grass	8,107	17,300	87	113	70
	Subtotal	8,107	17,300	87	113	70
Non-traditional						
	Steel	1,665	2,053	10	23.3	10
	Plastic	157	187	1	19.1	0
	Other (cloth, etc.)	336	376	2	11.9	0
	Subtotal	2,158	2,616	13	21.2	10
Total		10,265	19,916	100	94.0	80

and household appliances were distributed by international aid organisations to all communities affected in the aftermath of the calamity.

Even so, in comparison to traditional artefacts, the bulk of modern assets still remain low. In 2004, modern assets comprised bicycles (41%), radios (82% of all households), watches (61% of all households), agricultural tools and basic household appliances. Bicycles, watches and to a lesser extent radios are regarded as status symbols, as their acquisition is innately linked to cash. This is in stark contrast to household appliances and agricultural tools that are often exchanged for rice or plantains. Bicycles and watches are exclusively owned by male residents and, so I have learnt during my stay, not likely to be shared with women. On one occasion I tried to teach Andrea how to ride a bicycle, a light-hearted event that was a source of amusement for both of us. Upon telling her husband, however, I was met with considerable irritation. '*Ham pajqui pen'in*' was his sudden response which literally translates into 'not for women'. Radios are usually owned by male family heads or in fewer cases older sons who engage in wage labour. They come in different shapes and sizes and have become an essential item for every household. While agricultural tools include axes, spades, chisels, the indispensable machetes and more recently rice seeders (owned by 56% of all households in 2004), household appliances range from aluminium pots, plates, knives and spoons. Yet whenever there is a shortage of individual plates or spoons, plantain leaves and sticky green plantain peels aptly serve as the perfect substitute for the missing modern utensils.

Non-traditional communal assets in Campo Bello are very limited and consist of school furniture, a solar panel and a communal telephone. The latter two were installed in an attempt to develop the community and connect it to the outside world; both had, however, ceased to function when I arrived in the field. As a general observation, people do not show great appreciation for communal belongings and usually do not take great care of them.

4.3.3 Local Material Flows

Following the two system boundary distinctions, let us now move on to examine the material flows the social system of Campo Bello exchanges with its natural environment (domestic extraction and ouputs) as well as to other social systems (imports and exports). I will present the data following the inflows/outflows logic, thereby commencing the material flow analysis, by first presenting the weight of materials appropriated from the domestic environment and second by looking at imported materials from other social units. What follows is an in-depth analysis of nutritional flows and the use of water. The material flow balance then proceeds with the system's outflows. To this end, I will examine the wastes and emissions discarded into the domestic environment and then proceed to look at material exports to other social systems.

4.3.3.1 Domestic Extraction

Table 4.6 indicates the total weight of biomass appropriated from nature by the people of Campo Bello. Before entering into a detailed discussion of each of the categories, we find that timber extraction accounts for the highest biomass fraction, followed by agricultural harvests. Biomass appropriated from hunting and fishing activities makes up for 6% of the total weight extracted, followed by biomass gathered in the forest. Biomass extraction from livestock grazing barely shows, as in 2004, the people of Campo Bello owned only five heads of cattle (Table 4.7).

Agricultural extraction comprises the first category. Plantains make up for 45% of the bulk weight harvested, followed by rice that accounts for 35% of the total weight. This is not surprising since plantains can be harvested all year round and are an essential a staple in any Tsimane' household. Rice, on the contrary, can only be harvested once a year and rice stocks become depleted rapidly shortly after the harvest season through marketing. While both cultigens are favoured cash crops, it is the rice which has gained greater market importance in recent years. Manioc, on the other hand, is generally less popular as a market crop since, in the absence of transport opportunities, the cultigens perish fairly quickly after harvesting. Over the course of 1 year, a total of 70 kg of manioc (that is 9% of the total weight) are extracted by each member of the community, making for abundant quantities of manioc beer. Manioc equally serves as a staple for the community's small livestock population. Small patches of maize are found in almost all garden sites and people value their relative short production period of 3 months and easy handling. For the year 2004, total maize extraction accounted for 7% of the total extracted biomass.

Table 4.6 Total domestic biomass extraction, Campo Bello, 2004

	Total/a (t)	Total per cap·a (t)	%
Agri-horticulture	174.28	0.75	41
Hunting and fishing	27.35	0.12	6
Gathering	21.71	0.09	5
Woods and canes	198.13	0.86	46
Livestock grazing	6.30	0.03	2
Total	427.77	1.85	

Table 4.7 Domestic extraction from agriculture, Campo Bello, 2004

	Total/a (t)	Total per cap·a (t)	%
Plantains	78.16	0.34	45
Rice	61.45	0.27	35
Maize	12.74	0.06	7
Manioc	16.18	0.07	9
Fruits	2.81	0.01	2
Vegetables	2.94	0.01	2
Total	174.28	0.75	

The Tsimane' relish diversity and plant many crop varieties (see also Piland 1991; Vadez et al. 2004). My own field records contain six rice varieties, five varieties of plantains, four maize varieties and six varieties of manioc. On average though, most people cultivate only two or three varieties of each. For the inexperienced eye it was quite difficult to make out the distinctions of the different crop varieties, particularly when it is some morphological peculiarity of the root or negligible colouring differences. When asked about the reason for multi-cropping, the most common answers were '*por costumbre*' (it's a habit) and '*nos gusta*' (because we like it). Arguably though, from a purely culinary point of view, the multiplication of seed and crop varieties only marginally increases the range of flavours. During food sampling, men would recognise by taste alone only a very low percentage of the varieties brought into the house. Another explanation is brought forward by Descola (1996: 166), whose research on Achuar crop varieties led him to conclude that cultivating a rich pallet of plant species demonstrates a woman's competence as a gardener (the Achuar garden is entirely a woman's domain) and shows the very special bond she establishes with the crops she cultivates.

Contrary to the extraction of field crops, fruit and vegetable harvests from kitchen gardens happen on a much lower scale, accounting for 4% only of total annual biomass extraction. Still, the Tsimane' palate relishes the agreeable diversity of fruits as they come into season, among which mango, pineapple, cacao and watermelons figure most prominently. Bits of fruit peel and fleshy tissue scattered around the housing area, a welcomed treat for poultry foraging in the yard, are a reminder of the lush but relatively short fruit season from November to February every year. After snacking on a fruit, individuals meticulously guard fruit seeds and kernels and conserve them for re-cultivation. As opposed to the relative short fruit season, most vegetables can be harvested for longer periods. Most households have at least a few tomato or chilli shrubs, avocado trees or onions near their dwelling. Nonetheless, onions and tomatoes are not particularly valued for their gastronomic virtues and time and again I came across tomatoes being left to rot on the ground. An examination of the population's annual nutritional intakes tells a similar story; the consumption of onions accounts for a mere 2%, while tomatoes figure even lower at 1%. In 2004, under the auspices of a small-scale development project, the cultivation of organic beans was introduced with the aim of enriching peoples' daily diet, while also serving as supplementary chicken feed. Upon return to the field in 2006, however, only three families still planted this crop, at the last outposts of their kitchen gardens.

Adding to the staples obtained from agro-horticulture, additional food enters the social system via foraging and fishing activities. Meat returns from hunting account for 26% of total biomass extraction from these activities. While lacking a pronounced hunting season, there still exist fluctuations throughout the year that have an effect on hunting returns. Starting in January and February, prolonged periods of heavy rain have a dreadful impact on hunting, for they keep the men at home and drive animals away. With the dry season picking up again in March, hunting endeavours resume and peak in May and June when animals are fattest (due to the abundance of wild fruit varieties during the rainy season) and considered most tasty.

Hunting returns drop again in low-water season, extending from August to November, when animals travel further in search of waterholes. This period largely coincides with an increasing availability of fish, making for such bountiful catches that many families prefer the easier task of fishing.

We have learnt in the previous chapter that hunting is more than a mere food procurement strategy; it stands for manliness and requires the knowledge of, and adherence to, many taboos. But like in our society, the men of Campo Bello vary widely in their interests and abilities for hunting. Some are keen and fearless hunters who frequently strike out on large-scale hunting expeditions in the early morning mist. Due to their hunting expertise, they are well respected in the community, as they always fulfil their wives' craving for meat. While most husbands seem to accomplish their duties, the only adult male who never expressed any interest in hunting was the *profe*; he preferred to spend his leisure time in a neighbouring community where he eagerly tended his agricultural fields. There is a lot of evidence that game resources in Campo Bello are becoming ever more depleted. Residents frequently refer to their 'empty forests' and lament deplorable hunting returns as compared to the past. Owing to this growing game exhaustion in and around Campo Bello, hunters are increasingly forced to widen their hunting radius to the surrounding savannah region or deeper into the forest. This is why local hunters mostly prey on small animals such as peccaries, rodents, primates and birds. Large mammals like deer, tapir or peccary, while desirable, are becoming increasingly rare. Only once I recall catching sight of a tapir roasting deliciously over the open fire.

Shotguns or rifles are the most frequently used hunting devices in Campo Bello and in 2004, 34 men were in possession of such weapons. There is general consensus about the effectiveness of shotguns over other hunting armoury, converting them into highly solicited items to be obtained from the market. Only Simon, one of the elderly informants, spoke with nostalgia of the old days when forests were full of big game and blamed the use of the shotgun for scaring them off. In addition to his shotgun, a hunter rarely leaves home without arrows of various types in order to be fully equipped for his trip into the forest. These include knob-headed arrows for smaller birds and jagged-edged ones for large fowl or monkeys. Also, boys and some men take a slingshot with them that serves to stalk birds on high treetops. From an early age on boys already learn how to manufacture their own slingshots and indefatigably practice with mud balls on tiny living targets such as frogs, beetles or other insects. As a general observation, hunters are particularly careful that their weaponry is not subjected to the slightest damage. When not in use, a shotgun is lashed upright to one of the wooden house posts, whilst bows and arrows are cautiously stored in the roof, where they are kept out of direct sunlight in order not to warp the wood.

While hunting is a serious man's business, fishing tends to be more of a family affair. Fishing provides a reliable source of protein and takes a central place in the daily village life of Campo Bello. As listed in Table 4.8, almost 15 t of fish were extracted in 2004, accounting for the highest proportion of biomass appropriation besides agriculture. Similarly to hunting, fishing in Campo Bello takes place virtually all year, yet seasonal and climatic fluctuations have an effect upon fishing returns.

Table 4.8 Forest products extracted domestically, Campo Bello, 2004

	Total/a (t)	Total per cap·a (t)	%
Hunting	12.79	0.06	26
Fishing	14.56	0.06	30
Tree crops	5.06	0.02	10
Honey, turtle eggs, etc.	0.33	0.00	0
Palm leaves	7.20	0.03	15
Fibres	9.12	0.04	18
Total	49.06	0.21	

When water levels are high in January and February, the Rio Maniqui can become a dangerous site for fishing and families resort to smaller streams and a nearby lagoon about half a day's walk from the community centre. When water levels recede, fishing in the Rio Maniqui picks up again since this period coincides with the annual upstream migration of various fish species between March and May every year. People say that the unpleasant cold spell, extending from June to July each year, keeps fish levels low as they 'freeze' in the chilly river water. Moving into August, low water levels invite pleasant fishing trips and returns are highest during this period, which roughly lasts around late November. My own records reveal 15 different fish species that entered Campo Bello during my stay with the Tsimane', however, undoubtedly others escaped my attention (see Pérez 2001 for a detailed account of fish species among the Tsimane'). This range provides for a highly varied and reliable source of protein; from the most common *vonej* (*Prochilodus nigricans*) to the unpretentious *sona'are'* (*Pseudoplatystoma fasciatum*) via the more desirable *cajsare'* (*Salminus maxillosus*). As to their technology, locals mainly fish with hook and line, arrows or machetes, although, fishing nets are gaining particular importance among the local fishermen. In 2004, a total of 21 male household members were in possession of synthetic fishing nets. People occasionally use old mosquito nets as, contrary to hook and line fishing, they facilitate relatively large catches on a single trip. As previously mentioned, I did not observe any *barbasco* fishing parties and there is general consent about the adverse effects caused by this technique. Interestingly, I found some of the older informants lamenting today's paucity of tasty fish in Campo Bello as compared to former times, different reasons were given for this phenomenon. Pedro, for example, would blame the use of nylon fish nets for scaring away the fish. Others asserted that many fish were being poisoned by the motorboats (mostly river traders and park rangers) passing through the local waters. Still others were convinced that *barbasco* fishing is still practiced in secret and in their view is the reason why some fish species have ceased to exist.

Hunting and fishing activities are commonly combined with the gathering of forest foods. In light of the vast agricultural diversity and relative abundance of game and fish, gathering wild foods is largely a supplementary activity. The extraction of edible tree crops, for example, accounts for just 10%, as compared to the much higher biomass extraction obtained from hunting or fishing. However, the list of edible forest foods is enormous and their significance within the yearly production cycle should not be underestimated. The fruit season runs from November to February and during this time I frequently encountered entire families scouring the forest for tasty fruits. People keep a close eye on the time each plant comes into fruit and even children can tell exactly the different moments of the productive cycle of a single plant. The most common tree crops recorded on my daily walks through the Tsimane' forest were *motacú*, *chonta* (*Bactris* sp.), mango, sweet limes and *Rheedia achachairu*. During this verdant season, there is hardly anyone who does not nibble on some fruit or another in the course of the day. The gathering of honey comes into season in late August and extends until October. Not particularly a major undertaking, it accounts for less than 1% of the system's total biomass harvest. Still, honey is undeniably regarded as a special treat as it satisfies the peoples' sweet tooth, and that especially goes for the female palate. The honey season coincides with the season for turtle eggs, which are highly solicited as a trading currency in exchange for market products. In terms of quantity, however, turtle egg harvests remain almost negligible making up for less than half the amount of wild honey brought into the community.

Aside from food, the most important items extracted from Campo Bello's domestic environment are palm leaves and fibres, accounting for 15% and 18%, respectively. From the 7.2 t of palm leaves gathered in 2004, more than two-thirds were extracted either for the construction of new huts or for the purpose of roof maintenance before the onset of the rainy season. The remainder constitutes for its main part uva grass, a plant that is extracted by women throughout the year, but especially during the dry season when water levels are low and the river banks where this plant grows are easy to get to. It is used for weaving floor mats and other useful household items like fans or storage bags. For manufacturing mats, just the upper stem is cut and after leaving the young leaf shoots to dry in the sun for a day or two, about twenty-five leaves are then woven into sturdy floor coverings. A certain tree bark that can be extracted from different plant species (e.g., *Cochlospermum* sp.) in the area is used to fasten the mats. It usually takes a woman about 6 h to manufacture a floor mat, however, some women are more expert weavers than others. Barvarita, for example, would complete such a task in less than 4 h. Skilled hunters are also fond of arrow reed (*Gynerium saccharoides*), a cousin of uva grass, as they use the plants' dried culms to manufacture shafts for arrows.

The yearly extraction of fibres (*cajñere*) has been calculated at 40 kg per capita. Fibre harvesting activities peak between late August and early October, as they largely constitute the gathering of fibres from the balsa seed pod, a fruit with a texture similar to that of cotton. It used to be no more than a minor activity in Campo Bello as gathering activities were limited to using the woolly flesh for stuffing mattresses and pillows. Yet the advent of an industrial mattress production project in San Borja also affected gathering dynamics in the village. No longer for subsistence only, women have started to engage quite substantially in the marketing of *cajñere* as the project has become quite a lucrative business. For each *arroba* (11.4 kg) of this soft tissue marketed, they are paid 20 Bolivianos. Other wild fibres are extracted from bark and used in lashings, the main method by which all kinds of artefacts are held together.

Table 4.9 Wood and timber extraction, Campo Bello, 2004

	Total/a (t)	Total per cap·a (t)	%
Firewood	162.00	0.70	82
Construction wood	16.84	0.07	8
Uva grass	1.23	0.01	1
Non-subsistence wood	18.06	0.08	9
Total	198.13	0.86	

Let us move on to examine the weight of the wood and cane resources extracted from the domestic environment. The use of firewood is enormous (at 82%) but hardly surprising considering that fuel wood is the sole energy source used for cooking. Fetching firewood, mainly a woman's task, constitutes the gathering of smaller logs from adjacent agricultural fields. While wood resources are generally abundant all year round, the only time when dry logs might be in short supply is just before the opening of new agricultural garden sites between July and September. People then resort to seeking dried branches along the riverbanks. As a general observation though, fields are never burnt down to the root, leaving scraps of dry logs lying around. I was informed that the perfect fire requires an average of four to eight logs to be laid out in a star pattern, with logs pointed towards the centre of the fire. Each piece should have a length of about 2–3 m and a diameter of approximately 10–20 cm; only then can the fireside be operational for about 3 days. Building and preserving a good fire is a true art, entailing a good grasp of the burning properties of different woods and how to contrive large logs and brushwood to attain the desired heat (Table 4.9).

As concerns the category of construction wood, lumber extraction accounts for 8% of the total wood harvest. This number comes from observing the erection of five new dwellings during field research.[18] Mahogany is the tree species most frequently exploited for house posts and beams in Campo Bello; at the time of research, more than two-thirds of all housing structures were made of this tree species. For the manufacture of canoes, on the other hand, local men prefer *Calophyllum brasiliensis* and cedar, that both provide a relatively soft surface that facilitates wood carving. Adding to these usages, wood provides the raw material for manufacturing a whole variety of every-day artefacts such as the wooden vat for manioc beer storage, the mortar and pestle, both prerequisites for all kinds of food processing, or the odd stool found in some Tsimane' households. Finally, some men are fabulous toy makers and like to surprise their offsprings with creatively carved toys like pushcarts or simple wheels. Uva grass is the main source for housing walls and serves to make the platform beds within the Tsimane' home more comfortable. Due to their light weight their extracted mass weight remains low, amounting to 1% only of the total wood extraction in Campo Bello. At various times of the year, some

[18]A total water content of 14% was assumed and taken account of in the weight calculations for all tree species.

Tsimane' families camp in temporary shelters on the river beaches in order to take advantage of abundant fish resources or turtle eggs. These simple shelter constructions are usually made of uva grass poles with a couple of *motacú* palm leaves used for roofing.

Logging for purposes other than subsistence is strictly prohibited in the territorial area of Campo Bello, a fact that relates to its geographical location within the buffer zone of the BBR. This is in fact rather contrary to some other Tsimane' settlements further upriver, where the long-term presence of legal and illegal individual loggers and logging firms has had an impact on deforestation of precious hardwood species. However, non-subsistence wood extraction, albeit small-scale and sporadic, does take place in Campo Bello and was calculated at 18 tons per year, a number that slightly exceeds the mass weight extraction of construction wood. In 2004, for example, the community agreed to fell more than two dozen trees and sell them to a logging firm in order to generate the cash needed for the annual village feast. The undertaking had a semi-official character and the Tsimane' Council was informed by the village leader. To round off the metabolic profile of domestic biomass harvest, the extraction of grass through livestock grazing accounts for the final fraction and, in quantitative and qualitative terms, reflects indeed its insignificant character.

4.3.3.2 Imports

What is striking is the small weight fraction of imports as compared to domestic biomass harvests; the material weight generated from imports is 16 times lower than that generated from domestic extraction (Table 4.10). Biomass accounts for the largest segment of imported goods (73%), followed by manufactured commodities such as portable radios, bicycles and household appliances (20%). The mineral section is minuscule (4%) and solely contains the import of salt. The only fossil fuel imported (3%) is kerosene, a necessary energy source for night lighting. Kerosene candles are mostly simple devices, often consisting of an old tin bottle that contains a wick. As to the biomass inflows, the import of livestock and human food was accounted for. Livestock trading from other social systems, to begin with, is not a particularly common activity as most livestock is reproduced within the system. In 2004, livestock imports over the course of the year comprised little more than three dozen head of poultry and a couple of pigs, thereby accounting for less than 2% of biomass imports. The remainder were food imports for human consumption.

Table 4.10 Total imports, Campo Bello, 2004

	Total/a (t)	Total per cap·a (t)	%
Biomass	19.76	0.09	73
Fossil fuels	0.74	0.00	3
Manufactured artefacts	4.45	0.02	20
Minerals	1.06	0.01	4
Total	26.01	0.12	

Within the food import category pasta noodles rank highest at 43%, followed by beef from the market that accounts for 18%. What this shows us is the growing trend that pasta noodles are gradually replacing rice as the main staple consumed in Campo Bello. Informants of all ages like the good taste of pasta noodles and when asked about their preferred staple, pasta would figure prominently on most peoples' food popularity scale. Beef is either purchased on the market or simply traded with nearby cattle ranchers. Bread and flour imports are becoming growingly popular (15%) and Tsimane' adults hardly ever return from a trip to San Borja without having purchased at least a tiny bag of soft bread rolls which are generously distributed upon their return home. The Tsimane' take to sweets or biscuits as special treats that are particularly cherished by children and women. Tinned fish, industrialised fats and condiments comprise the rather insignificant remainder of imported food stuffs and account for 5%, 4% and 3%, respectively. Yet looking at the big picture, imports account for less than a quarter of all foods consumed in Campo Bello.

Fossil fuel imports to Campo Bello are almost negligible and simply constitute small amounts of kerosene (very rarely also diesel) that are used for lighting the house after dusk. Bottles with kerosene are neatly stored in the roof of their dwellings and families are not likely to share it with others. It is a highly valued resource, probably partly related to the fact that it is not always available in the market town of San Borja. Every household owns at least one night lamp that uses around 1.5 l of kerosene per month in order to guarantee candlelit evenings. While BBR park rangers employ diesel-powered motorboats to move about the reserve, all villagers use hand-made canoes as their mode of transport. The noise of the motorboats frequently attracts the villagers to gather along the river shore and curiously watch the park rangers pass. Transport services to the near market town of San Borja are repeatedly offered and willingly accepted by many locals.

Manufactured artefacts make up a total of 20% of all imported items. I have split them into two subcategories, durables and consumables, with their sole distinction being their life span. As to the former, the import of bicycles, radios, kitchen utensils, shotguns and agricultural tools account for less than 10% of total imports of manufactured commodities. The large remainder falls within the other subcategory and contains all products that are consumed shortly after having been brought into the system. They include estimates on the weights for soap, clothes and more short-lived products such as cigarettes or school material (Table 4.11).

The last category 'minerals' comprises salt imports, totalling 4% of all imports. Salt is an important resource among the Tsimane' and is also embedded in their cosmology (see Chapter 3). With the historical salt lick *Pat'sene* losing its importance

Table 4.11 Manufactured artefacts imported to Campo Bello, 2004

	Total/a (t)	Total per cap·a (t)	%
Durables	0.33	0.001	7
Consumables	4.12	0.018	93
Total	4.45	0.019	

in the 1960s, salt started to be purchased from the market. Local families like to keep it in plastic bottles stored in woven bags that freely dangle from the rafters. A local household buys an average of 2 kg of salt per month, of which less than 1 kg serves directly as seasoning for cooking, whilst the remainder is used for processing game meat into *charqui*.

4.3.3.3 Nutritional Flows

We have already observed that the Tsimane' diet offers a wide variety of tastes and that their gastronomic life is far from repetitive. Dependent as they are on the bounty of nature, the Tsimane' enjoy a great diversity of food sources and despite seasonal fluctuations, food is generally abundant all-year round and people enjoy a constantly varying supply of delicious foods from their fields, gardens and the forest. Moreover, the families I have been with have a tendency to eat even tough, bony, dry and crunchy parts of the animals entering the household, thereby adding to the great diversity of domestic edibles. According to the study results, 83% of all foods consumed in Campo Bello are extracted domestically. While many foods are eaten raw, such as the tasty nibbles on frequent gathering trips, most of their daily diet, however, comprises starchy food staples that must be cooked (e.g., manioc and plantains).

Ellis (1996) has written extensively about the consumption of food and drink as a social activity and concludes that food sharing is an intimate act that normally takes place within the confines of the household. The consumption of beer, in contrast, tends to be a shared social event when people cluster together around the patio of someone's house, fervently awaiting the gourds of manioc beer to be passed around the circle. Things were somewhat comparable in Campo Bello where I also observed peoples' protection of food resources, while at the same time not many of my daily rounds were accomplished without my Tsimane' hosts offering me a taste of their strong brew.

Plantains are the most important household staple, accounting for 59% of all food consumption. They are available all year round and there exists a great variety of delicious plantain recipes. Quickly roasting peeled plantains directly in the coals imparts a different flavour (and adds a dose of ash) than baking them unpeeled in the embers of the fire for quick snacking. Likewise, plantains are also the main staple for a common dish called *jo'na*, a sticky soup made of chunks of meat or fish boiled up with a porridge of scraped green plantains. Also, plantains are frequently added to the production of manioc beer and some women are particularly renowned for preparing plantain *chicha*, a refreshing drink without alcohol that is a particular treat for children.

Rice accounts for the second most significant staple in Campo Bello. Contrary to plantains or manioc, rice production is exposed to seasonal fluctuations. Particularly during the lean pre-harvest period up to March, people time and again mention their desire for fresh rice and count the days until the first rice variety (usually the *noventa* rice variety that takes 3 months to mature) is due for harvest. When the harvest period then finally comes along, many families, sometimes lured into unfair trading mechanisms, sell large quantities of rice to river traders visiting the community. Vadez et al. (2005: 9) similarly argues that Tsimane' farmers do not take advantage of price fluctuations and sell most of their production when prices are at their lowest. It was somewhat staggering to observe that, time and again, a number of families traded rice for pasta noodles. People usually pointed to the nice taste of pasta noodles when being asked why pasta would be more solicited on their daily menus. Indeed, annual rice consumption makes up for merely 21% of all staples, as compared to the growing consumption of imported pasta noodles (currently at 17%). Also at village feasts, pasta noodles and purchased alcohol seems to be somewhat preferred over home-grown rice and home-made beer (see Byron 2003). These are examples of the changing status of foods. It seems that imported foodstuffs have gained a more prestigious status as compared to locally produced foods; the Tsimane' palate adjusting to these socio-cultural changes. Despite its lesser popularity as a staple crop, the consumption of manioc comes in a variety of ways. Either roasted, boiled, grated into a soup or simply dried and preserved as *chive* (a refreshing drink made of dried manioc), it accounts for 8% of all staple food consumption. Maize is consumed on the cob but generally ranks low on the scale of favourite dishes. While not particularly solicited in their solid form, both staples are however an indispensable ingredient for home-made beer. Adding some maize to the manioc beer apparently increases the alcohol content and thus makes for the ideal brew.

Fish ranks highest in the consumption of proteins and accounts for 44% of total meat and fish consumption. Often, fish are steamed in their own juice by wrapping them in banana leaves or cornhusks and placing them directly in the coals. In addition to fish, game meat plays an almost equally important role in the people's diet (36%). Since people repeatedly voiced their concerns about the scarcity of game meat at various times of the year, it was not surprising to see how they treated game meat as an enormously valuable commodity that only got dished out in small portions. Meats are preserved either by salting for long conservation or simply sun-dried or smoked immediately after a catch. Purchased beef is still not consumed in large quantities, accounting for 10% of total meat and fish consumption. Beef is regularly consumed communally at village feasts, while individual families often obtain beef in exchange for maize or plantains at nearby cattle farms. The only meat provided domestically is that of chickens and ducks (7%). Only in recent years have some families acquired a taste for tinned sardines purchased from the market (3%). Although food samples were carried out in the wet season that endows the people with succulent forest and garden fruits, the consumption of fruits and vegetables only accounts for 13% as compared to the overall diet. Bread, flour, sweets, lard and oil are all purchased from the market and are becoming increasingly integrated in local dietary patterns. Finally, I have estimated the preparation and consumption of six vats of homemade beer per week. People have different ways of preparing this brew and not everyone likes their beer to be too strongly fermented. Most Sundays, Paula would prepare refreshing non-alcoholic plantain beer with sugar, a special treat for her grandchildren. Other women like Anastasia or Dalinda, spent significant amounts of time preparing traditional manioc beer, which involves the spitting of well-chewed chunks of manioc into the beer pot, and were respected throughout the community for their aptitude.

However, a nutritional analysis would be incomplete if just weight was accounted for, since examining the nutritional content of the main staples can be just as instructive. To start with, the local diet is relatively high in carbohydrates. This is not unusual, as plantains and manioc, typical for many Amerindian societies, are high in carbohydrates and low in proteins. Whereas farm production is the main source of carbohydrates, it is still the forest which provides the Tsimane' with dietary protein in the form of fish and game meat (Byron 2003). According to nutritional findings published by Godoy et al. (2005a: 13), Tsimane' per capita protein intakes average 86.3 g/day. If we compare these with studies of other Amazonian societies we arrive at very high levels. In his nutritional study of the Yanomami, Lizot (1978 in Descola 1996: 315) calculated the maximum protein needs for a grown man at 27.4 g/day. Another source (Spedding 1975: 93) indicates adequate protein intakes at 55 g/day for agrarian systems. If we take the latter value, the average protein intake of the Tsimane' is 157% of the safe level. Moreover, since the Tsimane' traditional diet is exceptionally low in fats (except for fish), lard and oil from the nearby market town of San Borja are highly desirable commodities for all households in the community. Interestingly, I found that all households in my sample lived on a relatively similar diet, except for the teacher's family who deviated somewhat from the prevalent food norms. Contrary to other families, a much smaller fraction of their daily food consumption actually came from own productive activities. All game meat and almost all of the fish they consumed during my stay in the village had indeed been a gift from the community, a sign of respect for his teaching and leadership abilities. The family's strong preference for market foods such as noodles or sugar became obvious when Juanita, the *profe's* wife, invited me for a delicious fish soup made of tinned sardines, noodles and a whole variety of tasty herbs that she had purchased at the San Borja food market.

Upon my return to the community in 2006 in the immediate aftermath of a terrible river flood that had destroyed large field areas, things had changed somewhat. Once a week, people would now gather along the river shore expecting the food aid allocated to most communities in the area. With international funding, representatives of the local municipality would distribute food parcels containing several litres of oil, flour, rice and sugar. While lacking any quantitative data from that period, these food interventions undeniably led to – probably temporary – changes in local nutritional patterns. In the mornings, my neighbours would now invite me for large quantities of fried bread and baked plantains would be seasoned with heaps of sugar. This was indeed a practice hardly witnessed the year before.

4.3.3.4 Water Use

Seven wells are found on the community territory, all located within walking distance from households and shared among extended families. Before their installation in 2001, all household water was extracted from the nearby Rio Maniqui.

Table 4.12 Total water use, Campo Bello, 2004

	Average per cap·a (l)	Average per cap·d (l)	%
Within the household	2,960	8	35
At source	5,249	14	62
Agriculture	0,231	1	3
Total	8,440	23	100

But still today, there are households who make repeated use of both sources to obtain water and the reasons are twofold. Some people reject the water extracted from the river for cooking as it changes colour when heated. It turns into a brownish liquid, a fact that can be explained by its high iron content. At the time of research, other families resorted to the river as their well had simply ceased to function. As a general observation, for the families of Campo Bello, the river provides a vitally important ecological and social resource. It is what people navigate on, what they bathe in at dawn, where women wash their clothes, dishes and other utensils (half of all local households interviewed extract water from the river to wash their clothes) and where children enjoy playing. Fetching water is a daily chore carried out by both, women and children, who manually carry large jars from the sources back to the house where the water is stored for common usage (Table 4.12).

A breakdown of total water use illustrates that the water utilised within the household (for cooking, drinking and washing dishes) accounts for 35%, whereas nearly twice as much is consumed at the source (for bathing, washing clothes).[19] People are generally careful with the usage of water and regard it as a highly valuable resource. Children are reprimanded, for example, if too much water is thoughtlessly spilt on their way back from the source. The use of water within the household is relatively consistent throughout the year. When asked about the amount of water fetched on a daily basis, women can tell straight away how many buckets are required for the domestic chores ahead, with answers varying little in spatial (households located closer to water sources) and temporal (seasonal changes) terms.

Water consumption directly from the well, however, paints a different picture. Presumably, the provision of accessible ground water sources has added to an increase in water usage in Campo Bello. Basing my argument on narratives from the locals, I assume that the washing of clothes in Campo Bello has become a much more frequent domestic activity as compared to before. When I shared a well with Andrea, for example, almost every day would she place piles of clothing into a

[19] Calculations are based on the following estimates (all numbers per household): Within the household: (1) cooking: 16.5 l/d, (2) use for making beer 45 l each time (based on the assumption that around six households brew beer every week), (3) 13 l/d for drinking/cleaning, (4) 15.75 l/d for washing dishes, At the source: (1) 65 l/d for bathing, 225 l/week for washing clothes – both calculations are based on the assumption that half of the actual water use for both activities is taken from the river – this water fraction was not included in the water use balance. Agriculture: 50 l/week during the 6 month dry period.

wheel barrow and fill it up to the top with water before thoroughly scraping off the dirt. Lastly, small-scale irrigation makes up for barely 3% of total water use and is limited to the watering of horticultural house gardens during the dry season. As natural rains are copious during the other half of the year, there is no need for additional irrigation.

4.3.3.5 Outputs

Having looked at the inflows, nutrition and water use within the social system of Campo Bello, the chapter now carries on with the outflows – outputs onto the natural environment and exports to other social systems – in order to complete the metabolic profile of the local community. Outputs are those wastes and emissions that are excreted from the social system into the domestic environment. The calculation of these output flows entails an analysis of (1) bio-metabolic wastes of the human and livestock population (2) household wastes, deliberate disposals and losses and (3) emissions to air and water (Table 4.13).

Despite the existence of several steel latrines in the village, the Tsimane' commonly go to the adjacent forest to defecate. The sight of any excrement left by children or domestic animals causes angry reactions among the adult members of the household who quickly collect the faeces into a leaf and dispose of them in the garden. The only bodily substance that is expelled in public, seemingly without any constraints, is saliva. Women's saliva, for instance, is the key to the fermentation of manioc beer, and liberal use is made of it during its preparation. Due to the low livestock numbers their waste discharge remains largely insignificant. The few cattle in the village are left to graze on a natural pasture, and smaller animals forage freely around the house where they indiscriminately leave their dung that serves as a natural fertiliser for the variety of plants found in the kitchen garden. The second category of output flows comprises household wastes and deliberate disposals. The structure of this analysis has been inspired by Mayrhofer-Grünbühel (2004), whose remarkable biophysical study on the village of Nalang in rural Laos uses the same

Table 4.13 Wastes from human and livestock bio-metabolic profile, Campo Bello, 2004

	Total amount	Total/a (t)	Total per cap·a (t)
Humans	231	9.47	0.04
Cows	5	4.96	0.02
Dogs	54	2.13	0.01
Cats	39	0.51	0.00
Pigs	3	0.55	0.00
Monkeys	5	0.07	0.00
Poultry	498	6.54	0.03
Others	67	0.18	0.00
Total	902	24.41	0.11

Table 4.14 Household waste, deliberate disposals and post-harvest losses, Campo Bello, 2004

		Total/a (t)	Total per cap·a (t)
Consumables			
	Non-organic	1.23	0.01
Deliberate disposals			
	Seeds and stalks	2.52	0.01
Losses			
	Post-harvest	26.10	0.11
Total		29.85	0.13

subcategories that constitute wastes from consumables and cooking, deliberate disposals and finally post-harvest losses (Table 4.14).

Household waste from consumables (household appliances made of plastic, aluminium, tin and cloth) is a minor business in Campo Bello and comprises a mere 4% of total outputs excreted into the domestic environment. The Tsimane' of Campo Bello generally recycle plastic bottles, tins and old clothes for quite some time before discarding them for good. Families usually designate an area away from the house where the waste is accumulated until, once heaps are amassed, the remains are set on fire. Burning takes place occasionally and over the course of a year hardly exceeds three or four times. For the flow analysis, an annual average of 30 kg of non-organic waste per household was calculated. Waste occurring during food preparation is just as negligible; cooking is always done in the open, where foraging animals are all too keen to get their share of food scraps. Rice husks, for instance, a side product of the rice husking process, are a treat for chickens and are devoured very quickly. Bones, scales, skins, fruit and vegetable peel and shells are other wastes that accrue during food preparation. Most scraps, however, have a limited life span, as they are hastily gulped down by all sorts of animals foraging near the fireplace. Skins or shells from game animal, on the other hand, are carefully preserved as hunting charms by some of the families or put to other more practical uses within the household. Upon finishing a meal, remaining chunks of smoked meat are carefully placed in woollen bags dangling from the rooftop, while other leftovers like soups or stews are kept in pots for later consumption; all leftovers are thus placed well beyond the reach of hungry dogs. Taking this into consideration, it was decided not to account for any organic waste figures within the material flow balance.

Deliberate disposals account for the second most important subcategory and include the quantities of seeds and stalks re-cultivated in the fields and garden sites. Household wastes finally include post-harvest losses for which a loss rate of 15% of total agricultural harvests was calculated. Post-harvest storage methods may be described as fairly poor and men would time and again ask me to bring back some rat poison from my occasional trips to San Borja. This is not necessarily surprising, as it is the palm-leaf ceiling of every house, often alive with the busy sounds of rats and insects, where bags of rice and maize are stored. Similarly if not put to use, manioc perishes quickly, providing a feast for all kinds of vermin. Leaving the waste categories excreted to land, the third and last grouping comprises emissions

Table 4.15 CO_2 emissions to air, Campo Bello, 2004

		CO2 output/a (t)	Total CO2 output per cap·a (t)
Breathing			
	Humans	84.32	0.37
	Livestock	22.19	0.10
Combustion			
	Firewood	296.46	1.28
	Litter	0.10	0.00
Total		403.06	1.74

to air and water. As concerns emissions to air, the study proceeds to examine the amounts of carbon dioxide (CO_2) released into the atmosphere by human and animal breathing and, even more importantly, combustion processes (Table 4.15).

Seventy-four per cent of all CO_2 outputs of the community of Campo Bello are a direct result of the combustion of firewood. As we learnt earlier in this chapter, firewood is the only energy source used for cooking purposes and logs are frequently left smouldering for some hours until they are next put to use.[20] The combustion of litter, though accounting for a miniscule fraction, equally adds to local CO_2 outputs. Due to its insignificance in quantitative terms, CO_2 emissions resulting from the combustion of kerosene were not included in the calculation table. The same goes for emissions to water. Local families do not own any motor-driven boats nor engage in customary *barbasco* fishing that is reputed to contaminate the water. As far as the pollution of river water through the use of soap and detergents is concerned, contamination levels have not been accounted for in this flow balance.

Let us finally draw attention to combustion processes generated from the prevalent agricultural practice of slash-and-burn. While not directly forming part of the material flow chart, combustion processes generated from this tradition equally add to the local CO_2 emission profile. If we believe the numbers put forward by Naughton-Treves (2004), the burning of field cover in Campo Bello would produce an incredible total of 3.5 t of CO_2 emissions per year. Yet burning after all seems to be a necessary element within the local agricultural cycle in order to sustain the moderately fertile local ecosystem for crop production (see Piland 1991; Huanca 1999; Vadez et al. 2005).

4.3.3.6 Exports

Before examining exports from Campo Bello to other social systems, it should be noted that there also exist internal material flows between households (Table 4.16). This is especially the case with extended family clusters, where meat and fish

[20] Open burning and smouldering increases Methane (CH_4) emission rates from fuel combustion. CH_4 emissions, however, are negligent in Campo Bello (extremely low number of ruminants, absence of flooded rice paddy fields).

Table 4.16 Total exports, Campo Bello, 2004

	Total/a (t)	Total per cap·a (t)
Agricultural biomass	57.26	0.25
Livestock	0.25	0.00
Timber	18.06	0.08
Handicraft and NTFPs	10.53	0.05
Total	86.10	0.37

Table 4.17 Agricultural exports, Campo Bello, 2004

	Total exports/a (t)	% exported of Total DE
Plantain	17.58	22
Rice	32.14	52
Maize	5.76	45
Manioc	1.78	11
Total	57.26	

among other staples are exchanged freely. Yet aware of the constraints I was facing as a lone researcher, any attempts to seriously account for all transactions, many of them happening in parallel, would have been illusionary (Table 4.17).

Agriculture provides the bulk of exports from the system of Campo Bello. In 2004, about 32 t of rice, almost 18 t of plantains, and 6 t of maize were marketed, with manioc at the lower end of the export scale. Piland (1991) and Vadez et al. (2005) both contend that rice has come to be the main cash crop within the Tsimane' agricultural system. This is also reflected in Campo Bello, where over the course of the year 2004, a total of 52% of all rice produced in the village was traded on the market, followed closely by the export of maize (45%), plantains (22%) and manioc (11%). Adding these numbers, this leaves us with a total of 34% of all agricultural harvests exported to other social systems.

Exports are cyclical, depending largely on individual harvest seasons. Plantains are traded throughout the year and all local families enjoy access to mature plantains at almost all times of the year, with the only lull period being during the months of June and July. In 2004, an average of 1.5 plantain heads per household per week was exported to other social systems. Aside from plantains, rice heads the list of most important cash crops, and harvests are frequently sold immediately after the harvest period in April or May every year. Following the logic of the free market, however, as the regional market becomes saturated from March each year, prices are at their lowest and only start to pick up again in late June. In 2004, a Tsimane' household would sell on average 68 *arroba* that is 775.2 kg of rice. Second on the cash crop list comes maize, which is far less frequently marketed, with local households trading 12 *arroba* or 136.8 kg during the course of the year. Manioc figures lowest on the cash crop list, with a single household selling a mere 3.7 *arroba* or 42.2 kg throughout the annual cycle.

The popularity of some cash crops over others, however, has more to do with trading channels and arising marketing opportunities rather than price. Piland (1991: 98–99) identifies three different types of marketing transactions taking place among the Tsimane': direct selling in San Borja, bartering with nearby cattle farmers and through intermediary river traders. As concerns the latter, he distinguishes marketing activities with part-time and full-time itinerant traders and refers to these transactions as the most exploitative deals for the Tsimane'. It seems that little has changed since his study. During my time in Campo Bello, I came to know three different traders who, by canoe or motorboat, frequently visited the community during the rice harvest period in order to obtain recent rice harvests from the villagers. Families usually know of their arrival beforehand and, after having meticulously prepared various bags of rice, get ready for the transaction. Relations with outside traders differ, with some families eagerly expectant of the forthcoming visits, while others bewail their unpleasantness. I particularly recall one incident when Rafael had been provided with a small monetary credit by one of the river traders that was to be paid back by a certain amount of rice. When he could not provide the large quantities he had promised, the trader became angry and started to considerably pressurise Rafael to increase his rice harvests. He would return periodically to remind him of his accumulating debts, until 1 day, still lacking the necessary means to pay the apparent amount overdue, Rafael decided to 'escape' from this burden by running off and relocating elsewhere for several weeks.

Export channels for plantains, on the other hand, are slightly different and somewhat steadier all year round. Only rarely do Tsimane' families embark on a trip to San Borja without taking a couple of plantains with them to exchange or sell for food or medicine. Manioc and maize are either sold to traders or, in the majority of cases, bartered for other foods (mainly beef) with nearby mestizo farmers. These farmers use Tsimane' products either for their own dietary needs or to produce manioc flour which they later sell to traders or cattle ranchers. The peak exchange period for manioc extends from May to September, when manioc flour is highly solicited in the area as it serves as the basic ingredient for *chivé*, a refreshing drink that provides a welcomed break from the strenuous physical work during the opening of agricultural fields. Since manioc roots perish shortly after being harvested, they are hardly ever marketed directly in San Borja.

Maize and rice, while being the most exposed to fluctuations in vending cycles throughout the year, still appear the most lucrative cash crops in terms of monetary returns. For rice, monetary returns are highest during the pre-harvest season between September and January, when there is a general rice shortage in the area. Market prices for maize are more stable throughout the year, with lower returns during the main harvest season extending from April to August. Manioc, on the other hand, generates low monetary returns throughout the year varying between $0.72 and $0.84 per *arroba*. These numbers, however, are an average approximation as, aside from seasonal vending cycles monetary returns equally depend on the marketing channel opted for. Amounts of annual livestock exports are minuscule and largely restricted to the transport of chickens to San Borja in exchange for batteries or other small items for daily use. At times, Tsimane' boys who master the art of

Table 4.18 Handicraft and NTFP exports, Campo Bello, 2004

	Total exports/a (t)	% exported of Total DE
Balsa flower	9.12	87
Eggs and honey	0.16	1
Woven mats	1.25	12
Total	10.53	

Table 4.19 (Un)Sustainability indicators, Campo Bello, 2004

		Total/a (t)	Total per cap·a (t)
Input indicators			
	Domestic extraction	427.77	1.85
	Imports	26.00	0.11
	Domestic material input (DMI)	453.77	1.96
Consumption indicators			
	Domestic material consumption (DMC)	367.67	1.59
	Physical trade balance (PTB)	–60.10	–0.26
Output indicators			
	Exports	86.10	0.37
	Wastes	54.26	0.23
	Emissions	403.06	1.74
	Domestic processed output (DPO)	457.32	1.98
	Direct material output (DMO)	543.42	2.35

slingshots catch wild parrots which they often keep as pets or simply sell them in town. San Borja's annual celebration, taking place on the 10th of October every year, is a particularly popular time for trading differently sized and coloured parrots (Table 4.18).

While agricultural trading tends to be a serious man's business, women engage in selling seasonally gathered forest products. Balsa seed pods account for almost 90% of total NTFPs and are sold almost entirely in San Borja in return for cash. We have come to learn elsewhere that between late August and early October, Campo Bello is bustling with women processing balsa flowers as they strip off the hairy fluff and make it ready for transport to San Borja, where a recently set up mattress factory pays them around $2.80 per bag. When I resided among the Tsimane', Rosa was much admired for her unfailing perseverance and had covered an incredible distance in search of balsa trees. Honey and turtle eggs present other marketable products worth gathering. Highly valued for its medicinal traits, wild honey has remained a locally sought after produce for years. Woven floor mats, a typical woman's product, make up the third subcategory, providing for a total of 12% of all NTFP products. Mats are either sold to nearby *mestizo* farmers who require them for the processing and drying of manioc flour, or, due to their relatively low weight, taken directly to San Borja (Table 4.19).

4.3.3.7 (Un)Sustainabiltiy Indicators

Following the detailed description of each of the biophysical material flow categories, the aggregate indicators that follow give a general assessment of the environmental performance of economies at national, regional and local levels, revealing our system's pressures on its environment in two ways: in general terms, inflow and consumption indicators give an insight into the resource consumption strategies of the system. Outflow indicators, on the other hand, represent (1) the degree of exploitation of the system by other social systems and (2) the degree of environmental pressure caused by the system's livelihood strategy.

Domestic Material Input (DMI) measures the input of used materials into the economy; these are materials that are of economic value and are used in production and consumption activities. As befitting a traditional society, the bulk of inputs into Campo Bello are made up of domestically extracted agricultural crops. Only a small amount of biomass is imported in the form of processed food and some livestock (4% of all biomass inputs). Imported finished goods merely comprise 17% of all imports. Nevertheless, parallel to changing nutritional habits among the local population (growing preference for pasta noodles) and steady stock increases, imports will equally experience an incremental growth rate. In addition, the indicator Domestic Material Consumption (DMC) reveals 'how much a society is willing or able to produce or harvest in addition to its own domestic consumption' (Singh 2003: 119). To put it differently, both the DMI and DMC indicators would be the same if we have a really subsistent society with no market dependency. In the case of our local society, people only consume 81% of all material inputs into the system. The remainder is used to maintain trade relations with the outside world. Export trade is not a new phenomenon and enjoys quite a long tradition among the Tsimane'. Itinerant traders have long exchanged consumer goods for rice, plantains and forest products. There is much evidence, though, that exports have risen in recent years as the people of Campo Bello are increasingly becoming exposed to external influences. Another indicator is presented by the Physical Trade Balance (PTB), an indicator that measures the physical trade surplus or deficit of the socio-economic system under discussion. PTB equals imports minus exports and points to the degree of exploitation or use of the society by other social systems (i.e., resources are used elsewhere). As to Campo Bello, the negative value reveals higher exports than imports, a fact that in the long run leads to natural resource depletion. The negative indicator reveals yet another equally interesting problem, namely the system's level of market integration. Whereas all exports are low-cost and unrefined, the people obtain high-value, industrialized consumer goods in return. This condition is very unlikely to change in the near future as for the time being, no value-adding manufacture or food processing opportunities exist within the confines of the community.

On the outflow side, the indicator Domestic Processed Output (DPO) reveals the environmental pressure social systems exert on their local environment in the form of pollution. As Mayrhofer-Grünbühel (2004: 133–134) rightly points out, this indicator only enables an examination of aggregate mass weights. Owing to their different composition, wastes and emissions are highly variable in the different environmental impacts they may cause. A final environmental performance indicator is Domestic Material Output (DMO), constituting the sum of outputs plus exports to other social systems. It therefore portrays the total quantity of outputs, both to the environment and to other social units. As for Campo Bello, exports only account for 16% of all outputs, while the remainder is made up of wastes (10%) and CO_2 emissions (74%) to the environment. Finally, the biophysical analysis equally provides insights into the physical stock growth of the system. Stock accounts for 2004 and 2006 not only reveal a spatial extension in physical structures but, more importantly, an increase in mass weight of traditional and non-traditional building materials, accounting for 10.5% and 12%, respectively. But there are differences between these types of building materials. Traditional materials, on the one hand, may be recycled or simply be left to decay. For modern concrete structures, on the other hand, the only possible future would be weight stagnation or further growth. A comprehensive discussion on the present biophysical opportunities, constraints and future options will follow at the end of this chapter. For now, we shall leave the sphere of material flows and turn our attention to the second important domain that constitutes the concept of social metabolism: the flow of energy.

4.3.4 Local Energy Flows

As opposed to conventional energy balances, the energy flow method follows a more holistic approach as it calculates all energy inputs into the system (Fig. 4.7). This also includes those energy-rich materials not directly used for energy conversion processes (e.g., construction wood) as well as energetic inputs provided by human and animal labour. The latter is a particularly useful aspect in more traditional societies where this kind of labour frequently plays a significant role (see Smil 2006). Energy flow accounting therefore provides an important database for the derivation of a number of energy indicators that assess changes in energy use and consumption patterns. Adding to this, an energy flow balance also provides a measurement of the dependency upon external sources of energy (Haberl 2001).

4.3.4.1 Domestic Extraction

Not surprisingly, domestic energy extraction from biomass accounts for the major energy fraction mobilised in the village (Tables 4.20 and 4.21). A more detailed look though reveals that agricultural biomass, while being the major source of nutritional energy for the local people, only contains one third of all domestically extracted energy. Of the main agricultural staples, rice takes the lead, comprising 60% of all energy stored in agricultural crops, followed by plantains containing 23%. Maize, despite its smaller presence in the local diet, stores about 12% of all agricultural energy,[21] while manioc ranks lowest among the staples, providing for a

[21] This relates to the high water content of plantains (74%) and the low water content found in grains (14%) (FAO 2001).

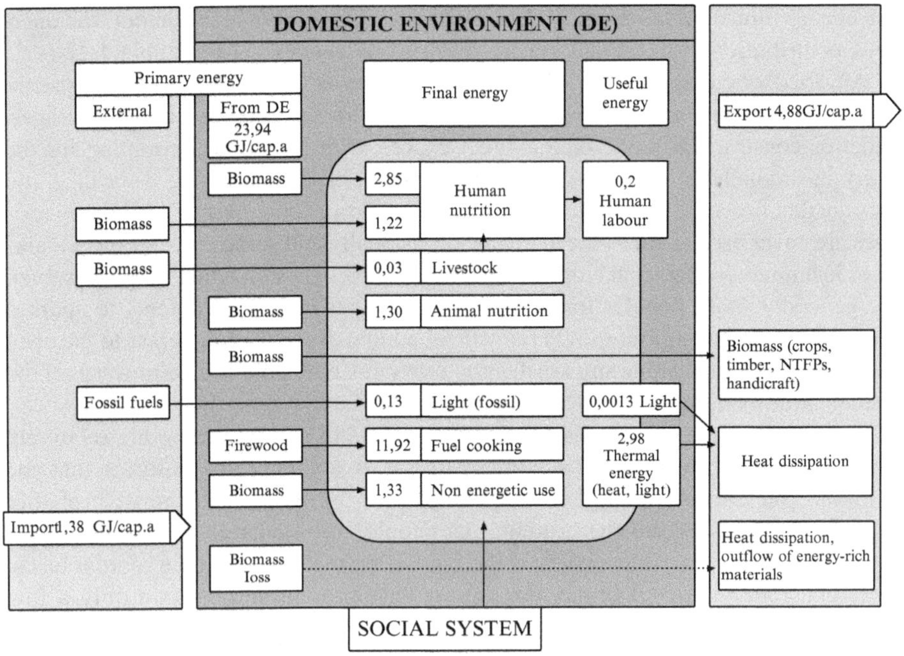

Fig. 4.7 Energy flows in Campo Bello, 2004

Table 4.20 Domestic energy extraction, Campo Bello, 2004

	Total/a (TJ)	Total per cap·a (GJ)	%
Agri-horticulture	1.61	6.96	29
Hunting and fishing	0.47	2.03	8
Timber extraction	3.37	14.57	61
Livestock grazing	0.09	0.39	2
Total	5.53	23.94	

Table 4.21 Domestic energy extraction from agriculture, Campo Bello, 2004

	Total/a (TJ)	Total per cap·a (GJ)	%
Plantains	0.38	1.63	23
Rice	0.97	4.19	60
Maize	0.20	0.87	12
Manioc	0.04	0.19	3
Fruits	0.01	0.04	1
Vegetables	0.01	0.04	1
Total	1.61	6.96	

3% energy return in the peoples' diet. Owing to their high water content, the energetic contribution of fruits and vegetables remains generally low (Table 4.22).

Air dried wood represents the largest energy fraction mobilised within the system. Of these resources, only 18% are domestically extracted for non-energetic usages, either as construction wood or for direct export, with firewood accounting for the enormous remainder of 82%. Again, this amount is hardly surprising if we take into account that firewood constitutes the sole energy source for cooking. While crop residues are never used as fuels, the Tsimane' occasionally add some extra straw or stalks from leguminous crops to the open fire. Dead tree branches or fallen twigs, however, are generally more popular instruments (due to low moisture content) to spark a proper fire on misty mornings. On rainy days, adding dry leaves and grass to the open fire generates considerable smoke, thereby providing a reliable source to ward off the sudden swarms of mosquitoes entering the house (Table 4.23).

The fact that energy returns from hunting, fishing and gathering are relatively low, somewhat confirms the statement that it is not only the calories that the Tsimane' seek from foraged foods. It is rather the good taste, most probably provided by the fats and the protein, that people are interested in. But why such low returns? One possible answer is the kind of animals hunted on a regular basis. Deer and rodents, for example, are almost pure protein and contain only a few percent of lipids. Interestingly, Smil (2006: 63–64) contends that low energy returns are often the reason why hunters embark on individual hunting endeavours rather than in time-consuming co-operative expeditions. Finally, the extraction of palm leaves and natural fibre constitute the largest energy yields within the range of gathering activities.

Table 4.22 Domestic energy extraction from wood and timber, Campo Bello, 2004

	Total/a (TJ)	Total per cap·a (GJ)	%
Firewood	2.75	11.92	82
Construction wood	0.31	1.33	9
Non-subsistent wood	0.31	1.33	9
Total	3.37	14.57	

Table 4.23 Domestic energy extraction from hunting, gathering and fishing, Campo Bello, 2004

	Total/a (TJ)	Total per cap·a (GJ)	%
Hunting	0.15	0.64	32
Fishing	0.07	0.30	15
Tree crops	0.01	0.06	3
Honey, fibres, etc.	0.12	0.52	25
Palm leaves	0.12	0.51	25
Total	0.47	2.03	

4.3.4.2 Imports

Imports comprise less than one tenth of the energy extracted domestically and largely consist of imported food items such as sugar, sweets, tinned fish, fats and pasta noodles (Table 4.24). The small remainder are fossil fuels, comprising kerosene for night lighting after dusk. There are, nevertheless, some indications that fossil fuel imports might increase in the future; Roman, for example, reiterated on various occasions that some of the local men were saving up for a diesel-engine motorboat.

4.3.4.3 Exports

Compared to the relatively small imported energy fraction, it is striking to note that energy exports amount to around 20% of all locally extracted energy (Table 4.25). Exports consist of biomass and involve agricultural products (61%), wood (27%), forest products (12%) and livestock (less than 1%). Singh (2003: 123) affirms that this one-way flow of nutrients leads to a break in the soil's nutrient cycle, putting in danger the regeneration of the local ecosystem.

Nevertheless, the river flooding in 2006 has, to a certain extent, brought about some changes to export. Even though no quantitative data was generated, exports must have fallen tremendously in 2006, as a high proportion of agricultural crops were destroyed by the river inundation. At the same time though, regular food aid interventions, albeit temporarily, increased the number of energy imports to the village (mainly energy-rich food stuffs such as refined sugars, oil and pasta noodles) and, in doing so, altered the energetic profile of the village.

Table 4.24 Imported energy, Campo Bello, 2004

	Total/a (TJ)	Total per cap·a (GJ)	%
Biomass	0.29	1.25	90
Fossil fuels	0.03	0.13	10
Total	0.32	1.38	

Table 4.25 Exported energy, Campo Bello, 2004

	Total/a (TJ)	Total per cap·a (GJ)	%
Agricultural biomass	0.69	2.97	61
Livestock	0.00	0.01	0
Timber	0.31	1.33	27
Handicraft and NTFPs	0.13	0.57	12
Total	1.13	4.88	

4.3.4.4 (Un)Sustainability Indicators

Following the same logic with material flows, the following energy flow indicators were derived (see Table 4.26).

Domestic Energy Input (DEI) measures the total amount of energy entering the social system. The largest part derives from domestically extracted biomass that accounts for 95% of all energy inputs into the socio-economic system. Within this biomass fraction, firewood used for daily cooking makes up 47% of total energy inputs. While at present, only a small amount of biomass is being imported, the figure has recently experienced an increase. How much energy does Campo Bello produce in addition to its own consumption? The answer can be given by looking at the indicator Domestic Energy Consumption (DEC). It reveals that 81% of all energy inputs into the village are consumed within its limits while the balance is traded with other social systems. In particular the marketing of relatively high-energy-yield grains is gaining importance among the local people.

4.3.4.5 Conversion Processes

Energy flows follow linear conversion processes in order to produce useful energy. In a first step, primary energy inputs, i.e., those inputs initially mobilised by a social system, are transformed into final energy. The final energy category contains all kinds of energy available before it is rendered useful by the final consumers (e.g., human nutrition). In a third and last stage, final energy is converted into useful energy, i.e., the kind of energy that can be mobilised to activate new sources of energy (e.g., through human labour) (see Haberl 2001) (Table 4.27).

As to final energy inputs, by far the largest portion is made up of human nutrition. Disaggregating the yearly data on human nutrition, the local people have an average daily nutritional intake of 2,664 kcal. In term of quantity we can presume that the Tsimane' of Campo Bello are well-fed, since the minimum food requirements for Bolivia, as recommended by the FAO (2001), are 2,500 kcal per day. In terms of energetic returns, imported food items account for 1.22 GJ per person per year, that is a total of 30% of all nutritional energy intakes. Livestock nutrition constitutes only one third of human food consumption. This is not surprising when bearing in mind that in terms of weight, people outnumber the local animal stock by almost 1:4. Animals forage freely around the house and are only occasionally fed with rice husks, maize and meat. A general observation is that animal husbandry in Campo Bello is rather ineffective. While it is true that merely 5% of total biomass extraction goes into animal feed, people consume an exceptionally low amount of livestock. It seems that animals are primarily appreciated as a kind of trading currency in exchange for consumer goods, rather than a mere food source.[22]

[22]Singh (2003: 121) points out that a traditional system with efficient animal husbandry shows an energetic return of around 10%.

Table 4.26 Energy flow indicators, Campo Bello, 2004

	DE (GJ/c·a.)	Total DE (GJ/a.)	Imports (GJ/c·a.)	Total Imports (GJ/a)	Exports (GJ/c·a.)	Total Exports (GJ/a)	DEI (GJ/c·a.)	Total DEI (GJ/a)	DEC (GJ/c·a.)	Total DEC (GJ/a)
Biomass	23.94	5,530.14	1.25	288.75	4.88	1,127.28	25.19	5,818.89	20.31	4,691.61
Fossil fuels			0.13	30.03			0.13	30.03	0.13	30.03
Total	23.94	5,530.14	1.38	318.78	4.88	1,127.28	25.32	5,848.92	20.44	4,721.64

Table 4.27 Conversion processes in Campo Bello, 2004

Final energy			Useful energy		
	GJ/per cap·a	Total GJ/a		GJ/per cap·a	Total GJ/a
Human nutrition (biomass)	4.07	940.17	Human work	0.20	46.20
Livestock nutrition (biomass)	1.30	300.30			
Process heat (biomass)	11.92	2,753.52	Process heat (cooking)	2.98	688.38
Light (fossil)	0.13	30.03	Light	0.0013	0.30
Total	17.42	4,024.02		3.18	734.88

What is striking is the fact that energy expenditure on food preparation is more than twice as high as the direct energy intake by humans and livestock. Fuel wood is the only source for cooking and it is not unusual to use the open fire all day. Especially when manioc beer is prepared, the ingredients are left to simmer in big aluminium pots for several hours before being removed and processed further. Also, fireplaces are used at any time of the day to smoke freshly caught fish and game meat. It seems rather wasteful to use fireplaces in the open, as they operate with low energetic efficiency. Presently however, there are no other more energy-efficient technologies in place. It should be recalled that the village does not suffer shortages of fuel wood as wood and timber resources near individual homes are abundant all year round. Kerosene for night time lighting is used by all households in the village. Candles are often very simple, often being old tin bottles that contain a wick. About 1.5 l of kerosene is required to assure a household's night time lighting for an entire month.

With respect to useful energy, by far the largest fraction (93%) is due to the process heat generated from firewood.[23] Together with the light generated with fossil fuels, it is part of the 'exosomatic' (Boyden 1992) energy flow of the society, as it does not flow through the human body. Only human work is used within the system and accounts for a mere total of 6% of useful energy delivered in the community. For the time being, no mechanical energy is being produced due to the lack of motorboats or other vehicles. As modes of transport, some men have bicycles while the vast majority simply walk on foot.

On an interesting final footnote, Ellen (1982: 119–120) asserts that populations culturally develop strategies that contribute to the efficient use of energy. In the case of the Tsimane' of Campo Bello, these strategies may include the trading of surplus resources from the forest or fields for high-energy-yield foodstuffs (e.g., refined sugars, pasta noodles). Strategies also include the assignment of daily recurring activities, such as fetching water, to children for whom it is energetically less

[23]Sources on the efficiency generated by open fires diverge between previous researchers conducting studies at the local level. Whereas Haberl (2002: 74) suggests an efficiency rate of 20% (taken from Dutt and Ravindranath 1993), Singh (2003: 122) points to an efficiency ratio of 25%. For the purpose of this study, the latter value was opted for.

expensive. Also, fields are most often located near the individual dwellings so as to reduce energy costs when harvesting crops and get firewood. Moreover several plantain saplings, an indispensable staple in daily village life, are planted within the confines of the house clearings in order to save long trips to the field. Finally, while there are many reasons for the Tsimane' to build their houses in the immediate vicinity of a river, this location undoubtedly saves the Tsimane' from carrying bulky produce over long distances as canoes readily await the heavy load for the ensuing transport to San Borja.

4.3.5 Local Colonizing Strategies

The colonization concept constitutes the second set of relations listed in Fig. 2.1. When people exchange material and energy with their natural environment, they cannot do so without changing certain parameters of the ecosystems they interact with. In light of this, the following pages describe the three different agricultural production systems in Campo Bello: kitchen gardens, agricultural fields and fallows. In order to calculate the intensity of colonization, the latter part of this section entails the empirical analysis of HANPP.

In Chapter 3 we have already introduced, at least on the surface, the traditional slash-and burn agriculture of the Tsimane' production system and briefly discussed general cultivation cycles, division of labour and their use of fallows. Building on this information, we now pursue a more in-depth analysis of their land use practices, which can be distinguished into (1) kitchen gardens, (2) agricultural fields and (3) fallows.

4.3.5.1 Kitchen Gardens

Rather small in size (on average 0.13 ha), the kitchen garden encircles the Tsimane' house like a belt. At first sight, kitchen gardens do not appear very impressive to an outsider, supposedly because in many cases they do not form a neatly laid out, separate entity. The edges of the house yard are generally a favourite spot for fruit tree species (like papaya, mango, avocado or guava), while a wide assortment of medicinal herbs, tobacco, peppers, onions, watermelons, peanuts or cowpeas, cotton, vegetable dyes and gourd plants can be found spread around the patio. Banana trees are placed around the perimeter of the garden site, thereby somewhat demarcating the family's housing territory. Some plants, like sunflowers or for the Tsimane' palate bland papaya, have an additional function as animal feed (see also Piland 1991) as poultry forages freely around the house.

The kitchen garden is predominantly a female space where men rarely venture. It is the women who are in charge of weeding and watering, although these activities do not necessarily demand a great deal of time and effort. According to my interview records, only about half of the respondents regularly weed their gardens with a machete or a scuffling hoe. Irrigation is equally undemanding as, at least during the wet season, natural rains are abundant. Only prolonged dry spells impel women to water their plants, albeit occasionally. This is in fact also reflected in the community's water consumption profile, as merely 3% of all water usage within the system is used for the irrigation of kitchen gardens. Men hardly engage in the garden work; it seems that their share is largely restricted to plucking a couple of tobacco leaves or checking on the growth of arrow cane. Descola (1996: 169) emphasises the fact that fish poisons must be tended by men, for if these were touched by women, they would lose their efficiency. As for the Tsimane', fish poisons are also regarded a man's cultivar, but, at least in Campo Bello, are not attributed with the same harmful consequences if tended by women.

The size and, more importantly, the opulence of a kitchen garden is intricately linked to two aspects: mobility and time. Since younger couples tend to be more mobile than older householders, their kitchen gardens are inclined to be smaller and less diverse. The size and verdure of a kitchen garden consequently depends strongly on the length of time a family has lived on one site. It may be thus deduced that the longer a house site is occupied, the more it develops into an increasingly complex agro-forestry unit. This entails dense vegetation layers on the upper levels that provide a barrier against heavy rains that wash away the soil, and a ground carpet extending into a patchy medley of interweaving watermelons, squashes or sweet potatoes. Kitchen gardens are also a site of experimentation with new species and varieties and women take veritable pleasure to prove their agronomic virtuosity upon their fist successful harvests. Seed stocks are either purchased from the market, or brought back from the adjacent forest. Rosa, one of the eldest residents in the community, was much admired for her lush garden that always provided some pleasant shade on my daily rounds. Her garden hosts more than 40 different varieties of medicinal plants, fruit trees, annual and perennial crops, tuber crops, decorative plants and cultivars for animal feed. But not all households own kitchen gardens (only two-thirds were counted in 2004), but share them with their extended families who often live within the confines of the same house clearing. In the informal system of Tsimane' property rights, ownership of kitchen gardens still resides with the family who previously lived there. People always return to the old gardens in order to harvest mature fruits or vegetables.

4.3.5.2 Agricultural Fields

Slash-and-burn agriculture undeniably represents the central pillar of the local production system. Pérez Diez (1983), Piland (1991), Huanca (1999) have all published extensively on Tsimane' agricultural practices and their observations were of great help to complement my empirical field records.

Choosing the Site The flowering of the *o'ba* tree in April marks the onset of the Tsimane' annual agricultural cycle. This time largely coincides with the fading stages of the heavy rice harvest period and Tsimane' men slowly commence their search for suitable land for a new agricultural field. The choice of a proper site is not solely dictated by agronomic criteria, but more so depends on criteria like distance from home, strategic location and the kind of plant species found there, in particular *motacú* palm trees that can be used for roofing. In 2004, except for Ansermo's field that was well tucked away in the depth of the forest, all other sites were located within a 1 km radius of the farmers' homes. Only two farmers had erected a temporary shelter on the margins of their newly opened fields, where they would stay when strenuous work in the dry summer heat took its toll. The agronomic potential, however, also influences an individual's choice and aside from soil criteria, a flat, well-drained location on an alluvial terrace above flood level makes for a good horticultural site. The fact that most fields in Campo Bello are located near the Rio Maniqui on relatively dry river terraces equally facilitates general material transportation. The size of a future field depends on the ambition and aptitude of the farmer and when on site, he appoints certain distinctive trees as border markers. Once a site is selected, the village head is informed and the word spreads quickly among the community members about his choice. It was rather interesting to note that with the exception of two brothers-in-law, the Tsimane' do not cut the forest adjacent to each other. According to Huanca (1999: 52) this technique accelerates forest regeneration as plots are interspersed with natural forest galleries which function as natural windbreakers.

One of the strategic decisions a Tsimane' farmer has to take is whether to cut primary or secondary forest (fallow), a choice that demands finding a middle ground between labour expenditure and land productivity. In 2004, notably more farmers (66%) voluntarily opted for the opening of fields from secondary forest, thereby seemingly compromising higher productivity levels for less labour. Cutting primary forest requires more labour input at the clearing stage, but labour investments tend to drop enormously at subsequent weeding periods. If a general pattern was to be established, then younger farmers are prone to open a higher number and larger-sized fields from primary forest as compared to their older fellow residents. Also, younger generations tend to keep secondary growth fallows for a shorter time before activating their site again. The age range of re-cultivated secondary forest fallows spans from one to 15 years (Table 4.28).

Clearing and Burning The time when a Tsimane' farmer starts clearing trees primarily depends on whether he has chosen primary forest or secondary forest as a field site. As primary forest contains dense trees and lianas, clearing usually begins in late May or June. Clearing of secondary forest requires a smaller amount of labour and less time to dry, and so clearing may be delayed until late August or September.

Table 4.28 Number and size of agricultural fields opened in 2004, Campo Bello

	New fields opened in 2004	Median age of fallow	Total ha	Total ha/per cap.	%
Primary forest	15		15.52	0.07	34
Secondary forest (fallow)	50	4.92	30.32	0.13	66
Total	65	4.92	45.84	0.20	

Fields are prepared in three phases and first of all require farmers to clear the thick undergrowth with a machete, leaving larger trees for later. The last stage then consists of trimming the felled trees. Clearing the underbrush especially with primary forest is no small undertaking, as it forms a tangle of shrubs, saplings, bushes, and tree ferns. The main danger in brush clearing for the mostly barefoot men is the risk of disturbing a poisonous snake, harmful insects, a wasp nest hanging in the tree branches or simply sharp thorns. Contrary to other indigenous groups in the area, the Tsimane' never pile leftover debris from undergrowth clearing but frequently leave the branches scattered evenly throughout the field site to dry for several weeks until the seasonal winds begin (Huanca 1999: 54). Some 3 days to a week later, the felling of larger trees commences. Not all trees are systematically felled but deliberately left standing since they provide shade for ground plants, on the one hand, and are esteemed highly for their practical use values, on the other hand. Certain non-edible species spared clearing are for example balsa wood or palm trees that are both indispensable for house building. The Tsimane' apply a certain notching technique also found with other Amerindian societies, the Achuar Indians, for instance (Descola 1996). In a first step, this practice entails notching all smaller trees and leave them somewhat dangling until larger trees are felled, causing all trees to tumble together. As felling trees is a particularly strenuous physically demanding activity during the land preparation stage, it is the sole purview of adult men and even adolescent boys are generally prohibited from joining in. Collective clearings are not necessarily a common endeavour among the Tsimane' of Campo Bello, although this would relieve the work-load on each individual labourer quite substantially.

The subsequent burning takes place once all debris has completely dried. Burning is considered a delicate matter, since getting the best result requires a good intuition on the local weather conditions. A field needs to dry for 1–3 months, depending on the combination between various days of sunshine and steady winds. Burning should ideally take place just before the onset of the rains. Often though, people start to get nervous about the right time and burn their sites prematurely. When this occurs, field owners envisage a not so bright future in terms of time and energy, as more labour will have to go into weeding. Likewise, the ash produced from a good burn raises the pH levels of the soil and improves its nutrient holding capacity (Sanchez 1976). Burning is preferably done on a day when there is a light breeze and matches are used to set fire to the scattered debris directly. The Tsimane' are very mindful not to light any fires close to a stand of trees that have been spared clearing. In terms of sexual labour division, burning is the sole activity during the land preparation processes (except for the occasional clearing of light undergrowth) that is equally performed by women.

Planting After burning, the cultivation of crops takes place in a gradual manner. The planting of rice especially, is distributed over quite a few weeks, as often more than one variety is cultivated. The gradual timing not only synchronises the growth of crops, but also avoids labour peaks. Often, the whole family partakes in sowing rice, either using a dibble stick to punch small holes in the ground or employing the more advanced time and labour saving rice seeder technology. As with other agricultural tasks, the planting of rice is physically demanding; people in Campo Bello do not plant rice in rows but spread the rice seeds rather randomly across the field area. The planting of maize takes place in a similar fashion (although usually in rows), with several seeds being placed in the same hole. Often, two crops of maize are planted per year, one in May for seed and the more important crop shortly before the beginning of the wet season (see Piland 1991). The cultivation of manioc requires the use of a machete to dig a small furrow, insert manioc stems and cover it again to avoid soil erosion. Plantains are planted in rows and corms are usually transplanted from older fields into square holes.

Weeding Weeding has to be done regularly, and especially during the first year of cropping, before weeds root and flower to avoid future weed propagation. Depending on the crop, weeding usually takes place two to three times during the cropping period. Rice fields are weeded twice, once when the rice plants reach a height of about 20 cm and again after 3–4 months of cultivation when they are about 50 cm high. I was told that one special indigenous variety requires a mere third of the usual weeding time. Maize is weeded twice while manioc demands two to three times. With the exception of the scuffling hoe, which the locals use by scraping the surface off the soil and cutting off the top of weed plants, all weeding is done with the all-purpose machete. For plantains, for example, the people use a machete to cut back the weeds sprouting around the parent plant. By breaking up the hardened dirt at the base of the bundle, fresh weeds can be extracted in no time. Given the gradual decline in yield after the second or third year of field production, the struggle against weeds in the end demands more labour than it is worth in terms of the gradually poorer results. Since weeding efforts steadily decline, active fields are progressively transformed into fallows.

Harvesting While rice and maize are harvested during one harvest season, plantains are harvested all year round. The timing of the harvest equally depends on the variety of field crops. Harvesting rice, for example, is by far the most labour-intensive effort in the entire cropping cycle and, depending on the variety, takes place between February and May. The families in the village sow an average of two and three rice varieties, all of which are planted at different times. With the exception of *noventa*, a rice variety that, however little valued for its gastronomic virtues, is ready for harvest after merely 3 months, rice normally takes about 5–6 months to mature. The advantage of planting various varieties helps to overcome possible labour shortages within families. Rice harvests are done among a couple of family members who first cut the panicles with a sharp knife and stack them on a suitable spot on the field. After transporting the bulk back home, people walk on it to separate the grains from the panicle. In a last step, the single rice grains are spread out on a plastic sheet to dry in the sun, after which they are stored in bags. The maturing cycle for manioc takes around 7–8 months. Nonetheless, due to the cultivation of various species, most families are in a position to harvest manioc all year round. As manioc roots, once reaped, they rot fairly easily; they are usually harvested upon demand for beer production or when opportunities arise for sale to passing traders. Manioc is harvested by women who dig the tubers out of the earth with the use of machetes. For plantains it takes about 1 year to bear fruit; due to the ownership of

Fig. 4.8 A Tsimane' woman preparing beer

differently aged plants, however, harvesting is a continual process, whereby the producing plant is felled with a machete. Maize crops mature about 3 months after cultivation, yet they are often left to dry in the fields for much longer periods (Figs. 4.8 and 4.9).

4.3.5.3 Fallows

Fallows constitute the third agricultural production system in Campo Bello, the fundamental difference to agricultural fields being the amount of control the people exert over this ecological microcosm. Or to use sociometabolic jargon, while fields are considered 'colonized terrestrial ecosystems', meaning that are deliberately and continuously controlled (i.e., through the permanent control of weeds) in order to render them more useful socially, fallows, on the other hand, are dominated by natural vegetation patterns without (or with minimal) direct human control. Visits to fallows are rather sporadic and normally take place when the need arises for additional plantains, fish poisons or other valuable fibres of economic importance. Fallows are equally valued as a source of seed stocks (e.g., plantains or manioc) for newly opened fields.

Fallow sites are typically much smaller than active fields, as only some patches are maintained after the main harvest. As Huanca (1999: 68) asserts, 'there is no

Fig. 4.9 A Tsimane' woman weaving a floor mat

deliberate transition from cultivated land to secondary forest nor are there fixed marks to indicate when a certain swidden plot passes the *cúm* category'. This observation correlates with other areas in the Amazon, where in the indigenous perception fallows are not abandoned (see Dufour 1990). The local people maintain their ownership of the fallows for practically as long as they are cultivated. The longer a family remain in a community, the more fallows it owns. Not surprisingly, my records reveal that older, more established local families are in possession of more fallows than younger families. In the Tsimane' property right system, fallows can be passed down from older to younger family members.

4.3.5.4 Human Appropriation of Net Primary Production (HANPP)

After this detailed account of the local land use systems, we are now ready to measure the intensity of the peoples' agricultural practices in Campo Bello. This is done with the calculation of HANPP that entails a comparison of ecosystem patterns and processes that would be expected without human intervention, with those energy harvests presently observable in Campo Bello (see Vitousek et al. 1986; Fischer-Kowalski and Haberl 2007).

NPP_0

Since Campo Bello presents a relatively young community comprising fairly dispersed and small-scale human settlement areas, generating those land use categories that would prevail in the absence of any human intervention was not an overly difficult undertaking (Table 4.29).

Albeit scant, there exists some historical data on what the specific area would have looked like before the arrival of first settlers. In topographical terms, the area is flat and annual precipitation rates amount to 1,800–2,500 mm. Humid and dry savannahs are found in the village that are both inapt for agricultural purposes. There are some literary sources (GTZ 1997; CIDDEBENI 2003) which suggest that the primary forest found within the confines of the community can be classified as tropical moist deciduous forest, for which the coefficients offered by Ajtay et al. (1979), Haberl (1995) and Mayrhofer-Grünbühel (2004) were used in order to calculate the NPP_0 accordingly. The figure derived amounts to 244 TJ, representing the total potential energy in biomass produced by the local ecosystem.

NPP_{act}

47 of the entire village area is still covered by primary forest that can, in line with Amazonian forest categorisation, be classified as tropical moist deciduous forest (Table 4.30). With an altitude below 250 m, ecological formations are dominated

Table 4.29 Potential NPP in Campo Bello, 2004

	Area (ha)	% of territory	Aboveground NPP (MJ/m²·a)
Primary forest	491.00	80	48.16
Water body and beach	69.50	11	0.30
Dry savannah	36.15	6	8.05
Humid savannah	18.35	3	24.01
Total	615.00		

Table 4.30 Actual land use in Campo Bello, 2004

	Area (ha)	% of territory	Aboveground NPP [MJ/m²·a]
Primary forest	289.75	47	34.80 + 48.16
Secondary forest	133.79	22	11.58
Agricultural area	65.97	11	9.00
Water body and beach	69.50	11	0.30
Dry savannah	36.15	6	8.05
Humid savannah	18.35	3	24.01
Housing area (+ open space)	1.49	0	2.88
Total	615.00		

by evergreen seasonal forest. Vegetation types are varied as species and adapt to the degree of flooding. There are four strata in which the tallest trees reach 55 m in height (e.g., *Hura crepitans, Ceiba pentandra, Copaifera reticulata*), the co-dominant species grow up to 43 m in height and trees with thinner trunks grow to 15–30 m in height (*Attalea phalerata, Astrocaryum muru muru*). The last stratum is defined by the sotoforest, which is much less dense with undergrowth and contains various smaller tree species (Holdrigde 1975 in Huanca 1999: 41). These forest areas are of great importance for the local people as they provide a hunting ground for numerous animals. Similarly, the forest hosts a vast range of tree crops, palm leaves, honey and other fibres that are all vital for peoples' subsistence needs. Due to its location within the buffer zone of the BBR, there is a ban on timber cutting for purposes other than subsistence.

Plant species found in secondary forest are similar to primary forest, however less dense and more degraded due to prior human agricultural intervention. Patches of secondary forest can often be found on river terraces where land has previously been used for agriculture. On repeated occasions I was informed by elderly residents that about two or three decades ago, there hardly existed any secondary forest in the community. Today, land of this forest type makes up for 22% of the whole village area with a tendency to further increase in the near future, since villagers generally prefer secondary forest for the opening of new agricultural fields.

The agricultural area of Campo Bello accounts for 11% of the total land area. It comprises newly opened and older fields that are still under production. Fields are usually found on alluvial river terraces, more recently though, people tend to open them further away from the river shore, a measure of protection against recurring floods. Plantains and manioc, especially, do not withstand too much soil humidity. Agricultural field areas are therefore dispersed around the entire village area, in between primary and secondary forest patches. Cropping cycles vary but the general tendency is to plant rice during the first year of cultivation, followed by maize or manioc in the second and third years of production. A field is abandoned selectively and by stages, since there are substantial differences in the time it takes the different species and varieties to mature, as well as the degree to which the cultigens resist takeover by the natural vegetation. The cessation of weeding rapidly leaves the garden looking like a fallow site, but still allows the occasional harvesting of manioc or plantains for another year or two.

For a length of approximately 3.5 km, the Rio Maniqui winds its course through the territory of Campo Bello. During the dry season, beach areas almost double and become an important place for gathering turtle eggs. They also provide for temporary shelters during fishing trips when water levels are low. A total of 11% of Campo Bello's surface area is covered with water and beach areas. Locally known as '*pampa*', dry and humid savannah areas are a relatively widespread ecological formation in the Beni lowlands and their origins are still much debated upon (Hanagarth and Sarmiento 1990; Pouilly and Beck 1994). It appears, however, that their formation is almost certainly a result of the Pleistocene Epoch in the department of Beni. The humid savannah of the region is seasonally flooded, and combined with its loamy texture, makes it unsuitable for agriculture.

Human settlement areas account for less than 1% of the territory's land cover. As indicated on the village map (Fig. 4.2), individual dwellings are dotted here and there along the river; extended families usually clustering in close vicinity to each other. After the river flooding in early 2006, some of the dwellings that were severely affected by the high water levels were abandoned and houses reconstructed further inside the forest area. House yards are kept clean of weeds and regularly swept, the bare soil providing not only the cooking area for mothers, but also serves as a playground for children. Wooden benches or stools are scattered around the yard and passing neighbours are often invited to join beer gatherings or simply relax in the shade of a sheltering tree. This type of land cover also includes the communally owned football field outside the school, as well as the large patio behind the school building that serves as a location for communal happenings, such as village feasts. Using the coefficients provided by literature, the NPP_{act} has been calculated at 159 TJ. In this light, the local system of Campo Bello produces around two-thirds of the biomass energy that could potentially be produced by the natural system without human intervention.

NPP_h

Harvest, for our purpose, is the proportion of biomass reaped for socio-economic purposes (Table 4.31). This includes both, the biomass that serves to cover the local system's subsistence needs, as well as the energy released through the burning of forest cover.[24] The total amount of NPP_h accounts for nine TJ, representing an annual biomass harvest of almost 4% of potential energy production and 6% of actual NPP production in the ecosystem. A closer look at the individual figures though, reveals that the largest fraction (3.5 TJ) of energy is actually destroyed through the use of fire. The energy appropriated directly through the harvest of agricultural crops, on the other hand, accounts for half the amount, that is 1.8 TJ.

Table 4.31 NPP_h in Campo Bello, 2004

	Total harvest (TJ)	%
Agricultural crops	1.8	19
Hunting and fishing	0.2	2
Gathering	0.3	3
Wood and timber	3.4	37
Livestock grazing	0.1	1
Burning of forest cover	3.5	38
Total	9.2	

[24] Though not directly beneficial to the social system, the burning of forest cover is included in the calculation as failing to do so would distort the NPP figure prevailing in the ecosystem.

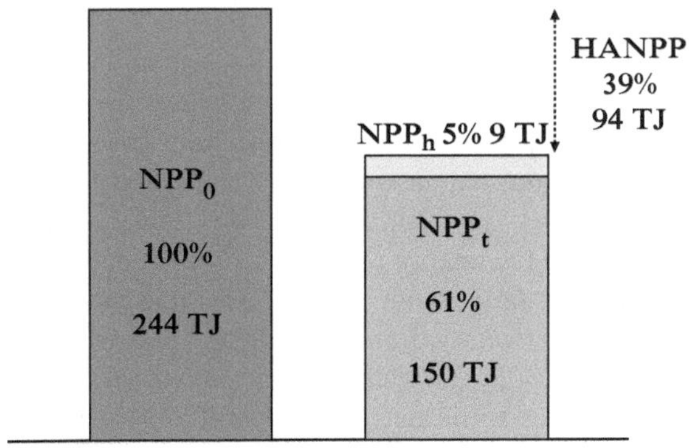

Fig. 4.10 Graphic illustration of HANPP, Campo Bello, 2004

The high biomass energy fraction of wood and timber harvests reflects the extensive use of firewood prevalent in daily village life.

HANPP

Having assessed the NPP_0 (potential vegetation), NPP_{act} (actual vegetation) and NPP_h (harvested biomass energy), the remaining energy in the ecosystem (NPP_t) is then calculated by subtracting the harvest from the actual vegetation ($NPP_{act} - NPP_h = NPP_t$) (Fig. 4.10). In do doing, we arrive at 150 TJ. HANPP is subsequently determined by deducting the remaining vegetation from the potential vegetation ($NPP_0 - NPP_t$ = HANPP). HANPP then accounts for 94 TJ that is 39% of the net primary production of the potential vegetation.

4.3.6 Possibilities, Constraints and Future Prospects

By taking a little distance from the details on material, energy and land use, we are to put the pieces together and interpret the empirical results. What are the system's possibilities and constraints in terms of production intensity and environmental performance? And considering a broader sociometabolic regime discussion, what are likely and unlikely future pathways for Campo Bello?

4.3.6.1 Primary Energy Use

According to Sieferle (2003), an agrarian sociometabolic regime obtains practically all primary energy from land use, either from productive cropland, grazing areas or forestry (see Fischer-Kowalski and Haberl 2007). The primary energy input into

Campo Bello is 25.32 GJ/cap/year, of which by far the largest fraction is biomass (99%). The remainder are fossil fuels that enter the system in the form of kerosene. Except for occasional periods of shortages, practically every household owns a night lamp that is fuelled by either kerosene or, in rarer cases, by diesel.

If we now compare our present case with the calculation of the yearly per capita primary energy input of past agrarian regimes, we derive a number that is much lower and in fact, closer to the energy profile of a sociometabolic hunter and gatherer regime. According to Krausmann et al. (2008), past agrarian regimes had a primary energy input of between 40 and 70 GJ/cap/year, while Fischer-Kowalski and Haberl (in Proinger 2005: 20) estimate an annual primary energy input for hunter and gatherer regimes to lie between 10 and 20 GJ/cap/year. Firewood for producing heat accounts for the lion's share (11.92 GJ/cap/year) of primary energy use; a rather unsurprising result since firewood is the sole cooking source in the village. This is not likely to change in the near future, as there is plenty of firewood still available within the boundaries of Campo Bello and hence, energy expenditure for this activity remains at low levels. Also in terms of time investment, firewood collection only takes up 14 min of a person's day (a detailed time use account will be provided in Chapter 5) and can be fairly easily delegated to children. The harvest of agricultural biomass from cropland (fields, fallows and kitchen gardens) makes up 27% (6.96 GJ/cap/year) of primary energy input, less than half of firewood. 2.85 GJ/cap/year of the agricultural biomass extracted from the domestic environment is directly used for human nutrition, whereby two-thirds derive from agro-horticulture (1.87 GJ/cap/year) and one third from foraging and fishing activities (0.97 GJ/cap/year). The latter confirms that it is not the calories the Tsimane' seek from foraged foods but good taste, which is most likely provided by protein, fats and general dietary diversity. Also, gathering walks are a pleasant pastime and a good excuse to break the monotony of everyday tasks.

4.3.6.2 Nutrition and Dependency

Hobbes (2005: 201) builds an insightful bridge between the concept of food security and material flows (in caloric units). She establishes a range of non-economic calorific food security indicators that are especially relevant for researching local systems in developing countries. For the purpose of my study, I have found two of these indicators particularly useful. The first concerns the degree of self-sufficiency (also Pfister 2003) that measures the amount of calories the system grows for its own need. For Campo Bello, calorific production (including exports) is about three times higher than the daily recommended caloric need for an average rural adult in the developing world. Looking at these numbers alone, we see that Campo Bello would be self-sufficient in the absence of food import channels. And this is not only the case in terms of calorific quantity, but also in terms of nutritional content. The second measures the potential degree of self-sufficiency which means the exclusion of exports. We still see that Campo Bello extracts about twice as many calories from its domestic environment as the daily caloric intake it requires. Calories from agriculture constitute the main fraction, accounting for almost 80% of the food energy extracted. These figures, however, only tell half the truth since they do not account for the culturally changing nutrition patterns we have observed in our system.

Imported biomass for human nutrition amounts to 1.22 GJ/cap/year and, while still lower than nutritional energy from domestically extracted biomass, the figure is likely to increase as eating patterns are changing. Byron (2003: 227) observed that among the Tsimane' imported food is increasingly valued socially. This becomes particularly visible with community feasts drawing closer, when cash quotas are collected from the villagers in order to purchase beef, bags of pasta noodles, sweets and alcohol. Smil (2006: 130–131) strongly advocates that dietary transitions happen in all populations as they become more modern. These transitions, so he further asserts, are accompanied by a qualitative shift in cereal consumption, from whole grains to highly milled products. Likewise, one observes a rising intake of refined sugars that are almost unknown in traditional societies, where sweetness merely comes from fruits and honey. Befitting Smil's assertions, intakes of refined sugars account for 4% of all food energy in Campo Bello, while traditional sweetness is derived from honey (late August to October), mango, watermelon and other minor fruits. Equally important is the gradual replacement of traditional rice with imported pasta noodles. The food sample method in 2004 revealed that rice accounted for 17% of total calorie intakes, while pasta noodle intakes equal 13%.

What follows is that the village is becoming increasingly dependent upon imported foodstuffs. This makes them less resilient to natural hazards and crop failures as well as to the exposure to externalities. If we look at the nutritional changes currently taking place, we can conclude that pasta noodles may function as an adequate substitute for rice in terms of iron, protein and energetic value. Bread and flour, however, do not serve as equal substitutes for domestically extracted carbohydrates from field starches. Among meat intakes, fish and imported beef are comparable in terms of energy provision. Nevertheless, fish contributes more natural fats to the local diet than beef does (see Byron 2003: 225–227). As rice provides for the main cash crop in the village, families tend to increase their rice sales in order to obtain processed foodstuffs and consumer goods in exchange. Interestingly, the availability of pasta noodles does not substitute the time and labour-intensive cultivation and harvesting of rice. The cash crop rice is rather used as some kind of 'currency' to pay for better-liked market substitutes.

4.3.6.3 Final and Useful Energy

The energetic metabolic profile of Campo Bello is further characterised by a low conversion efficiency of primary into final and useful energy. The primary energy input of 25.32 GJ/cap/year results in the availability of 17.42 GJ/cap/year and 3.18 GJ/cap/year of useful energy. Energy losses are enormous and entail a 31% loss from primary to final energy and an 82% loss from final to useful energy. Again, this is comparable to the energy metabolism of Austria in 1830 whose low energy conversion efficiency is expressed in similar figures: 38% and 80% losses, respectively (see Fischer-Kowalski and Haberl 2007: 226). Food for human and animal nutrition accounts for one-quarter of the community's total final energy supply, whereas human labour represents 6% of useful energy delivered. What is striking is the extremely low efficiency of animal husbandry. Animals are not used for work, but nor do they figure prominently on the peoples' food preference lists. Whereas animal feed accounts for 1.30 GJ/cap/year, merely 0.13 GJ/cap/year are actually consumed by humans, leaving 1.17 GJ/cap/year practically 'unused'. Yet again, while energy returns are low, so is the time invested in animal husbandry; just 5 min are invested daily per person. Animal husbandry involves only a limited range of tasks and for its most part is done by children.

The provision of heat relies on firewood, accounting for 68% of final energy used and more than 90% of the useful energy delivered. This is almost entirely used up in process heat, as space heating is hardly ever required in the Amazonian tropics. Again, this is fairly comparable to past agrarian regimes (Krausmann and Haberl 2002). As previously mentioned, open firesides are likely to retain a central place in the community life of Campo Bello. I base my assumption on two aspects: availability and costs. First, firewood is easy to obtain practically all year round. Secondly, a change to gas ovens would be a costly alternative for the families of Campo Bello – in terms of money as well as transport. What is in stark contrast to past regimes, however, is the, albeit miniscule, energetic throughput of fossil fuels for lighting, which account for less than 1% of the final energy used. For the time being, kerosene still remains the only exosomatic energy source in Campo Bello. Some families though increasingly express their desire for an engine-powered canoe in order to facilitate transport flows in and out of the community. While the purchase of a diesel boat would be unaffordable for a single family, the acquisition of a communal boat may be an option and was widely discussed when I last visited the community in 2008.

4.3.6.4 Material Use

The inherent energy limitations of their biomass-based metabolism have a direct bearing on Campo Bello's material consumption patterns. In relation to the community's domestic material inputs (DMI), biomass accounts for 98% of all inputs. Domestic material consumption (DMC) in Campo Bello amounts to 1.59 GJ/cap·a, with a total biomass share of 99%. Comparing my findings of a contemporary system with the historical metabolism of agrarian Austria in 1830 (Krausmann et al. (2008: 10) calculated a per capita material consumption ranging from somewhere between 5 and 6 t/cap/year, with the biomass fraction accounting for 87% – my figures bear more resemblance to that of a hunter and gatherer regime whose material consumption has been estimated at 1 t/cap/year (Fischer-Kowalski and Haberl 1997: 30). Aside from biomass, all other construction materials (concrete, steel, plastic) are imported and still used in low quantities in terms of volumes per capita and area. In terms of land use, modern building structures take up even less than 10% of the whole built-up area in Campo Bello. In terms of weight, however, the use of imported material accounts for 43% of the total weight of infrastructure and is likely to increase. The influx of non-traditional building materials started about a decade ago in an attempt to develop the community through the construction of a school and a well system. If we compare data from 2004 and 2006, for example, we see a 12% increase in imported construction materials. The steep increase (94%) in artefact consumption between 2004 and 2006 can largely be attributed to external aid interventions that took place in the aftermath of a severe river flood in early 2006. The largest fraction is owing to the replacement of manufactured heavy-weight artefacts (above all canoes) that had fallen victim to the floods. The range of smaller household artefacts for every day use had increased by about 25%. This aid intervention was quite a novel experience for the families of Campo Bello with the *profe* playing a key role in mediating between the community and the municipal government.

A noteworthy feature is the relatively high export rate to other social systems. About 20% of domestically extracted materials are exported, leaving the local people to subsist on the remaining 80%. The negative physical trade balance (PTB) is equally revealing as it illustrates that more material resources are exported than imported. If we look at the type of exports to other social systems, we are confronted with a social system of low market integration. People export unprocessed and low-value biomass products (i.e., unpeeled rice) in return for processed foodstuffs and other consumables. As long as there are no more lucrative market opportunities, community members are forced to make use of the local natural resource base to meet growing market demands. As valuable local resources become ever more depleted, the system may in the long term run into a deadlock. This observation reveals that our system has been taken up by the economic globalisation process, both on the input side through the purchase of consumer goods and, while to a much lesser degree in terms of market value, also on the output side. If we now examine the level of incorporation as suggested by Hobbes (2005: 200–201), which involves a community's degree of involvement in outside markets (through agricultural production and extraction, in our case), we find that, on the input side, Campo Bello has a level of material incorporation that equals zero. This has to do with the fact that no agricultural inputs (except for small quantities of agricultural tools or the odd hunting device) are in fact imported.[25] Things are quite different on the output side, where the relatively high export of biomass results in a degree of incorporation of 0.20 (total exports/total DE). Again, as this incorporation constitutes largely low-priced agricultural biomass, exports will most likely augment in the near future, as ever more Tsimane' are lured into the growing range of market goods arriving daily at San Borja's airfield.

[25]Total imports of agricultural tools, shotguns, ammunition, etc. were calculated at 90 kg/year for the whole community.

4.3.6.5 Demography and Water Use

A look at local demographic trends reveals substantially high growth rates. At an annual population growth of 3.77%, Campo Bello's population density of 38 cap/km² is similar to the estimates put forward by Krausmann et al. (2008) who claim a population density of 30–40 cap/km² under agrarian regimes. As roughly one third of this growth was due to a positive migration balance and two-thirds to births exceeding deaths, there is evidence that within the near future, the village population will grow rapidly. While forest resources are gradually becoming more scarce (e.g., depletion of game) and population numbers are growing, people will have to intensify the use of their resources. Since the introduction of wells in 2001, the use of water has undergone changes. In 2004, 62% of all water consumption in the village took place directly at the groundwater source. People obtained their drinking water from there, bathed and washed clothes. Prior to installation, the river provided for all the local water requirements. Today, there is still a range of families reluctant to use water from the well for cooking purposes and various women still opt for the river to wash their clothes. Nonetheless, the consumption of water at source is likely to increase as people are losing their apprehension through awareness-raising capacity-building projects to improve health and hygiene.[26]

4.3.6.6 Colonization Intensity – HANPP

As we are dealing with understanding the society's environmental relations, HANPP provides an adequate indicator to account for changes in ecological energy availability resulting from human activities. In other words, HANPP indicates the pressure on land resources and serves as a useful indication as to future potentials of natural resource use in the system. Campo Bello can be characterised by a foraging horticultural system (see Vadez et al. 2005) that sells agricultural surplus on the market. Being a relatively young settlement, the village area is still endowed with large forest resources. In fact, 47% of all local forest is still primary forest, whereas 22% can be classified as secondary forest. Practising extensive shifting agriculture, fields account for 11% of the total land area within the village boundary. Owing to its extensive land use practices, we found a relatively low HANPP of 39%. Extensive farming practices are a strategy well adapted to regional ecological soil constraints. In contrast to more intensive techniques that lead to high yields often in detriment to the ecosystem, the people of Campo Bello accept comparatively lower returns. Harvest efficiency (NPP_h/HANPP) is therefore correspondingly low, accounting for 10%. For now, local farmers only harvest about a quarter (1.8 TJ) of the actually available NPP produced by the agricultural area (6 TJ). Hence, the

[26]In 2005, the village was integrated into an extensive foreign-funded health and hygiene project covering various Tsimane' communities along the Rio Maniqui. The entire project duration is 5 years.

people of Campo Bello still retain some potential as to enhance their NPP appropriation. One should also note that more than two-thirds of the total NPP stored in biomass is harvested through the burning of forest cover. Though this practice destroys rather than provides NPP for human appropriation, the locally practised slash-and-burn technique is quite crucial, since only through the use of fire are the soil nutrients necessary for the growth of field crops released into the soil. In a similar fashion, our calculations reveal that the extraction of forest resources compared to the actual NPP production accounts for merely 2%. With regard to the natural NPP production of grasslands, the grazing of cattle extracts a negligible amount only, accounting for around 1%.

If we finally compare the biophysical profile of Campo Bello against the 'ideal type' sociometabolic hunter and gatherer regime, the agrarian regime and the industrial regime, we find our community as a hybrid system that presents many characteristics typical of agrarian regimes, while other features show more parallels with the hunter and gatherer regime. In terms of energy and material inputs, for example, our system is closer to the metabolic profile of hunters and gatherers. Biomass inputs, for example, account for almost 99% of total inputs. Similarly, our system's average food intakes are comparable to calculations for past hunter and gatherer regimes (see Haberl 2002). As far as Campo Bello's economy and social features are concerned, its metabolic profile shows more similarities with agrarian regime patterns. Parallels can be drawn in terms of population density, low conversion efficiencies of primary energy, the substantial use of firewood and biomass exports. Adding to this, the colonization of terrestrial ecosystems typically presents one of the key distinctions between hunter and gatherers and agrarian regimes. What is clear from our analysis is that the Tsimane' have always been swidden horticulturalists with hunting and gathering complementing their economic profile, a very typical feature of South American rain forest societies. However, what has changed over the last decades is the increasing importance of agricultural production, which has led to the shortening of fallows and an intensification of land use. Finally, fossil fuels – a key characteristic of the industrial metabolic regime – have been used in the form of night-light fuelling only. The use of diesel for motorboats thus far exists outside the community only, although these usages also impact upon Campo Bello. In this respect, the use of motorboats eases general transport and mobility, thereby prompting a greater influx of foreign goods.

4.4 Conclusion

This chapter has taken us on an empirical journey through the biophysical landscape of Campo Bello. It set out with a qualitative introduction into the village setting of Campo Bello in an attempt to provide a well-grounded base for later biophysical interpretations. From the metabolic study that followed we have come to learn of some of the empirical challenges when conducting a material and energy flow analysis at the local level. My message was that for these kinds of empirical field situations, a great deal of cultural sensitivity, as well as creativity, is needed if we want do such research. The following sections established first a material flow balance illustrating the inflows, nutritional flows and outflows. Following the same logic, energy flows were accounted for. What is different from conventional energy balances is the fact that the energy flows calculated for the purpose of this study include all energy inputs into the system, including those materials not directly used for energy conversion processes. As the concerns for the material and energy exchanges diminished, territorial issues and the impact of land use began to develop. Although related to the former, the presentation of Campo Bello's different agricultural land use systems gave insights into the current impacts of the local peoples' intervention in natural processes. The calculation of HANPP provided a specific measurement of the local colonization intensity. What is a striking result is the low harvest efficiency, with more than two-thirds of the total NPP being harvested through the burning of forest cover. Moving beyond individual biophysical resource measurements, the biophysical discussion that followed revealed the possibilities, constraints and some hints as to the future prospects of Campo Bello. The notable features are the high population growth levels that have already offset certain land use intensification processes (e.g., the shortening of fallows, growing secondary forest areas). Moreover, Campo Bello is characterised by extremely low primary energy conversion efficiencies, substantial usages of firewood and relatively high levels of low value biomass exports. At the same time, consumer goods are progressively more becoming integrated in community life; trends show that this is especially true for market foods which are increasingly considered of higher social value than domestic foods. What shows is a growing market dependency on the input side, while on the output side and owing to the type of unrefined biomass that is exported, a far less degree of market incorporation can be observed. If growing consumer levels are to be sustained for a growing population, their option space is fairly limited. One alternative would be a further intensification of agricultural practices in order to generate more exports; another option entails the increased engagement in outside wage labour opportunities. Both decisions, however, automatically increase the demand for labour investments either on the whole community or just certain age and sex groups. Hence, it is the burden of increased work-loads we are now going to encompass in Chapter 5.

References

Ajtay, G. L., Ketner, P., & Duvigneaud, P. (1979). Terrestrial primary production and phytomass. In B. Bolin, E. T. Degens, S. Kempe & P. Ketner (Eds.), *The global carbon cycle* (pp. 129–182). Chichester, New York, Brisbane, Toronto: Wiley and Sons.

Bernard, H. R. (2006). *Research methods in anthropology: Qualitative and quantitative approaches* (4th ed.). USA: Altamira Press.

Boyden, S. (1992). *Biohistory, the interplay between human society and the biosphere.* Paris: UNESCO and Parthenon Publishing Group.

Byron, E. M. (2003). *Market integration and health: The impact of markets and acculturation of the self-perceived morbidity, diet, and nutritional status of the Tsimane' Amerindians of lowland Bolivia*. PhD thesis, University of Florida.

CIDDEBENI. (2003). Satellite Image.

CIMAR. (2001). *Local material and energy flow analysis (LMFA)* Task 2. Santa Cruz: CIMAR, March 2001.

Descola, P. (1996). *In the society of nature: A native ecology in Amazonia*. Cambridge: Cambridge University Press.

Dufour, D. (1990). Use of tropical rainforests by native Amazonians. *BioScience, 40*(9), 652–659.

Dutt, G. S., & Ravindranath, N. H. (1993). Bioenergy: Direct applications in cooking. In T. B. Johansson, H. Kelly, A. K. N. Reddy & R. H. Williams (Eds.), *Renewable energy, sources for fuels and electricity* (pp. 653–697). London: Island Press.

Ellen, R. (1982). *Environment, subsistence and system: The ecology of small-scale social formations*. Cambridge: Cambridge University Press.

Ellis, R. (1996). *A taste for movement: An exploration of the social ethics of the Tsimanes of lowland Bolivia*. PhD thesis, University of St. Andrews.

FAO. (2001). *World agriculture: Towards 2015/2030*. Earthscan: An FAO perspective. Food and Agricultural Organisation, London.

Fischer-Kowalski, M., & Haberl, H. (2007). Conceptualizing, observing and comparing socioecological transitions. In M. Fischer-Kowalski, H. Haberl (Eds.), *Socioecological transitions and global change: Trajectories of social metabolism and land use* (pp 13–62). Cheltenham, UK, Northampton, USA: Edward Elgar.

Fischer-Kowalski, M., Haberl, H. (1997). Staffwechsel und Kolonisierung: Ein universalhistorischer Bogen. In M. Fischer-Kowalski, H. Haberl, W. Hüttler, H. Payer, H. Schandl, V. Winiwarter & Zangerl-Weisz, H. (Eds.), *Gesellschaftlicher Stoffwechsel und Kolonisierung von Natur: Ein Versuch Sozialer Ökologie*. Amsterdam: Gordon & Breach Fakultas.

Godelier, M. (1986). *The mental and the material: Thought economy and society*. London: Blackwell Verso.

Godoy, R., Reyes-García, V., Vadez, V., Leonard, W. R., Huanca, T., & Bauchet, J. (2005). Human capital, wealth and nutrition in the Bolivian Amazon. *Economics and Human Biology, 3*, 139–162.

Godoy, R., Reyes-Garcia, V., Huanca, T., Leonard, W. R., Vadez, V., Valdés-Garcia, C., et al. (2005). Why do subsistence-level people join the market economy? Testing hypotheses of push and pull determinants in Bolivian Amazonia. *Journal of Anthropological Research, 61*, 157–178.

GTZ .(1997). Proyecto *Manejo de zonas de amortigación de areas protegidas en Bolivia*. Informe sobre Diagnósticos Rápidos Participativos (DRPs) en cinco comunidades Tsimanes en el marco de la elaboración de un plan de ordenamiento de la zona de amortigación sur externa de la reserva de la biosfera estación biológica del Beni, San Borja.

Haberl, H. (1995) *Menschliche Eingriffe in den natürlichen Energiefluß von Ökosystemen: Sozioökonomische Aneignung von Nettoprimärproduktion in den Bezirken Österreichs*. PhD thesis, University of Vienna.

Haberl, H. (2001). The energetic metabolism of societies, Part I: Accounting concepts. *Journal of Industrial Ecology, 5*(1), 11–33.

Haberl, H. (2002). Economy-wide energy flow accounting. In Schandl, H., Grünbühel, C. M., Haberl, H., & H. Weisz, (Eds.), *Handbook of physical accounting. Measuring bio-physical dimensions of socio-economic activities. MFA-EFA-HANPP* (pp. 28–44). Vienna: Federal Ministry of Agriculture and Forestry, Environment and Water Management.

Hanagarth, W., & Sarmiento, J. (1990). Reporte preliminar sobre la geoecología de la sabana de espíritu y sus alrededores (Llanos de Moxos, Departamento del Beni, Bolivia). *Ecología en Bolivia, 16*, 47–75.

Hobbes, M. (2005). Material flow accounting of rural communities: Principles and outcomes in South East Asia. *International Journal of Global Environmental Issues, 5*(3/4), 194–224.

Holdrigde, G. (1975). *Como saben los ancestros*. La Paz: Historia Bolivia.
Huanca, T. (1999). *Tsimane' indigenous knowledge. Swidden fallow management and conservation*. PhD thesis, University of Florida.
Krausmann, F., & Haberl, H. (2002). The process of industrialization from an energetic metabolism point of view. Socio-economic energy flows in Austria 1830–1995. *Ecological Economics, 41*, 177–201.
Krausmann, F., Fischer-Kowalski, M., Schandl, H., & Eisenmenger, N. (2008). The global socio-ecological transition: Past and present metabolic profiles and their future trajectories. *Journal of Industrial Ecology*, accepted for publication, vol. 12, Issues 5–6, pp. 637–656.
Lieth, H., & Whittaker, R. H. (1975). *Primary productivity of the biosphere*. Berlin, Heidelberg, New York: Springer.
Lizot, J. (1978). Economie primitive et subsistance: essai sur le travail et l'alimentation chez les Yanomani. *Libre*, 78(4), 69–113.
Matthews, E., Amman, C., Fischer-Kowalski, M., Bringezu, S., Hüttler, W., Kleijn, R., et al. (2000). *The weight of nations: Material outflows from industrial economies*. World Resources Institute: Washington DC.
Mayrhofer-Grünbühel, C. M. (2004). *Resource use systems of rural smallholders. An analysis of two Lao communities*. PhD thesis, University of Vienna.
Naughton-Treves, L. (2004). Deforestation and carbon emissions at tropical frontiers. A case-study from the Peruvian Amazon. *World Development, 32*(1), 173–190.
Netting, R. M. (1993). *Smallholders, householders. farm families and the ecology of intensive, sustainable agriculture*. Stanford: Stanford University Press.
Nordenskiöld, E. (2001 [1913]). *Exploraciones y aventuras en Sudamérica*. La Paz: APCOB.
Pastore, G., Giampetro, M., & Ji, L. (1999). Conventional and land-time budget analysis of rural villages in Hubei province, China. *Critical Review in Plant Sciences, 18*(3), 331–357.
Pérez, E. (2001). *Uso de la ictiofauna entre los Tsimane*. Universidad Nacional Mayor de San Andrés: B.S. thesis.
Pérez Diez, A. (1983). *Etnografía de los Chimane del oriente boliviano*. PhD thesis, University of Buenos Aires.
Pfister, F. (2003). *Resource potentials and limitations of a Nicaraguan agricultural region*, PhD thesis, ETH Nr. 15169.
Piland, R. A. (1991). *Traditional Chimane agriculture and its relation to soils of the Beni biosphere reserve*. Bolivia: Master thesis, University of Florida.
Polanyi, K. (1944). *The great transformation: The political and economic origins of our time*. Boston: Beacon Press.
Pouilly, M., & Beck, S. G. (1994). Geografía general. In M. Pouilly, S. G. Beck, M. Moraes & C. Ibañez (Eds.), *Diversidad biológica en la llanura de inundación del Río Mamoré: Importancia ecológica de la dínamcia fluvial* (pp. 13–26). Santa Cruz: Imprenta Sirena.
Proinger, J. (2005). *Arbeitszeit und Nachhaltige Entwicklung in Europa: Ausgleich von Produktivitätsgewinn in Zeit statt Geld?* Social Ecology Working Paper 77, April 2005.
Reyes-García, V. (2001). *Indigenous people, ethno-botanical knowledge and market economy: A case-study of the Tsimane' Amerindians in lowland Bolivia*. PhD thesis, University of Florida.
Sanchez, P. (1976). *Properties and management of soils in the tropics*. New York: Wiley-Interscience.
Schandl, H., Grünbühel, C. M., Haberl, H., & Weisz, H. (2002). *Handbook of physical accounting. Measuring bio-physical dimensions of socio-economic activities. MFA-EFA-HANPP.* Federal Ministry of Agriculture and Forestry, Environment and Water Management, Vienna.
Scheper-Hughes, N. (1983). Confronting problems of bias in feminist anthropology. In N. Scheper-Hughes (Ed.), *Women's studies* 10(Special issue) Routledge: Philadelphia.
Singh, S. J. (2003). *In the sea of influence: A world system perspective of the Nicobar islands*. Lund: Lund University.

Sieferle, R. P. (2003). Sustainability in a World History Perspective. In B. Benzing & B. Herrmann (Eds.), *Exploitation and Overexploitation in Societies Past and Present* (pp. 123–142). Münster: LIT.

Smil, V. (1992). *General energetics, energy in the biosphere and civilization.* New York: Wiley and Sons.

Smil, V. (2006). *Energy: Beginner's guide.* Oxford: Oneworld Publications.

Spedding C. R. W. (1975). *The biology of agricultural systems.* London, New York, San Francisco: Academic.

UNFCCC. (1998). *Handbook on energy sector: Fuel combustion.* Retrieved January 14, 2007, from United Nations Framework Convention on Climate Change: http://unfccc.int

Vadez, V., Reyes-Garcia, V., Godoy, R., Apaza, L. V., Byron, E., Huanca, T., et al. (2004). Does integration to the market threaten agricultural diversity? Panel and cross-sectional data from a horticultural-foraging society in the Bolivian Amazon. *Human Ecology, 32*(5), 635–646.

Vadez, V., Reyes-García, V., Huanca, T., & Leonard, W. R. (2005). *Cash cropping, farm technologies and deforestation: What are the connections?* Tsimane' Amazonian Panel Study Working Paper 18, 18th of December 2005.

Vitousek, P. M., Ehrlich, P. R., Ehrlich, A. H., & Matson, P. A. (1986). Human appropriation of the products of photosynthesis. *BioScience, 36*(6), 368–373.

Werthmann, K. (2004). *A field full of researchers. Fieldwork as a collective experience.* Working Paper 33, Department of Anthropology and African Studies, Johannes Gutenberg University, Mainz.

Wu Leung, W., & Flores, M. (1961). *Food composition table for use in Latin America.* Bethesda, MD: Interdepartmental Committee on Nutrition for National Defense.

Chapter 5
Time, Work, and Other Functions

Abstract This chapter introduces the third set of metabolic relations: the concept of functional time use. To this end, the first part of this chapter gives a brief overview of existing sociological and anthropological time use studies, followed by a description of the functional subsystems (person system, household system, economic system, and community system) and the field methods for data collection. The second part of this chapter entails the discussion of the empirical results and is divided into three subsections. The first section deals with an analysis of Campo Bello's potentially disposable labour time by age/sex group. The second subsection takes a more holistic approach and examines the number of person-hours invested in the reproduction of each of the subsystems. The third subsection looks at efficiencies and strategies to increase time productivity. Finally, a discussion on the possibilities, time squeeze and bottlenecks that Campo Bello is currently facing, draws the chapter to a close.

5.1 Introducing and Embedding the Use of Time

Thus far, we have learnt that the exchange of energy and materials between social and natural systems creates certain, though often unintended, impacts on the environment. The higher the energy and material requirements of a social system, the more inflows (e.g., natural resources) are needed, which, in turn, produce more outflows (e.g., wastes or emissions) on the environment. At the same time, social systems also deliberately intervene in natural systems in an attempt to reap higher metabolic returns. This illustrates the clear link between the first and the second set of metabolic relations, social metabolism and colonization. This chapter attempts to integrate the third set of relations within this MEFA frame: the concept of functional time use. For our analysis, its point of entry is provided through the direct link between colonization and human labour. In the absence of fossil fuel

L. Ringhofer, *Fishing, Foraging and Farming in the Bolivian Amazon: On a Local Society in Transition*, DOI 10.1007/978-90-481-3487-8_5,

technologies in historical societies, human work was the only way for a society to appropriate nature (e.g., through agriculture). These two aspects, colonization and human labour, are also interdependent: the more a natural system was exploited by human hands and brought into a state far from natural equilibrium, the more labour investment was needed for the system to remain in this desired state. And, to tease out this thought further: the more ecological parameters were altered beyond the state of natural equilibrium, the higher the burden on the environment. With the onset of historical industrialisation processes – by tapping into the 'subterranean forest' of non-renewable energy carriers (Sieferle 2001) – this inherent link between environmental costs and human labour time came to be removed.

However, even today, there are many communities that retain characteristics of such traditional societies. One such feature is the sole use of human labour when people interact with their natural environment. Campo Bello is one such community. Apart from this link between labour time and environmental burden, the concept of functional time use equally provides an insight into the burden of labour time on the social system or certain segments of the society. This is achieved by looking at the system's time/labour ratio, coupled with a detailed demographic analysis. What differs from other approaches to time use is the holistic approach that sees time as a limited resource, which needs to be invested throughout a 24-h period. One fraction of daily time needs to be expended on certain functions (e.g., sleeping or eating), that are necessary for an individual's basic reproduction, while the remainder is used according to social norms, economic necessities or simply individual preferences. By looking at the different functions that need to be sustained through the investment of human time at the communal level, we gain first insights into the constraints and opportunities a society has in its interaction with its environment (environmental costs) as well as its members among each other (social costs).

In light of this, the present chapter explores the concept of functional time use and applies it to the social system of Campo Bello. For this endeavour, I will first of all provide a brief overview of labour time from sociological and anthropological perspectives. The reasons for doing so are twofold. On the one hand, my empirical methods were inspired by most of these studies; this goes especially for the coding and classification schemes. At the same time, I also want to show the different approach to time use my study undertakes. This holds mainly for the theoretical approach that, based on a detailed demographic analysis, offers a macroscopic look at time use at the community level. Following the data collection process in the field, the second part of this chapter is dedicated to the discussion of empirical results. The analysis is divided into three subsections and starts by describing the socially disposable labour time at the community level according to the different age/sex groups. Secondly, I will look at the number of person-hours invested in each of the functional subsystems (person system, household system, economic system, and community system). The third subsection deals with efficiencies and looks at the system's strategies to increase the productivity of time. This chapter draws to a close with a lengthy discussion on the possibilities and time constraints Campo Bello is currently facing.

5.1.1 Time Use Studies Revisited

Time use studies enjoy quite a long tradition in the social sciences, as they provide a window on actual lifestyles, needs and preferences, allowing for a rich, objective and replicable basis on which to make empirical judgements. Early accounts of sociological time use were predominantly concerned with exploring the social conditions of the rising working class, for which the publications *How Working Men Spend their Time* (Bevans 1913) and *Round about a Pound a Week* (Pember-Reeves 1913) provide testimony. In the early 1930s, a whole new era of work/leisure studies was launched. Among the range of emerging time diary literature, *Time Budgets of Human Behaviour* (Sorokin and Berger 1939) probably figured most prominently as it provided several new insights into sociological and psychological stimuli for daily time use. Since the 1950s, the effects of longer working hours have become increasingly analysed through the lens of comparative time use data. Undoubtedly the most ambitious multi-country time use study was the Multinational Time Use Study directed by Alexander Szalai (1972) in the mid-1960s. His methodological approach provided a landmark for future surveys in terms of data collection and coding schemes (see Harvey and Pentland 1999). Since then, comparative national level studies have flourished, with Jonathan Gershuny's (2000) Multinational Time Budget Data Archive probably taking the centre stage. Today, we owe sociology a great deal for having methodologically advanced the area of time use research; the statistical time-budget tradition particularly, has been a fruitful endeavour in many respects and has since been adopted by other disciplines. Despite the general praise it has received, Carlstein (1982: 333) somewhat voiced his critique on sociological time use studies, asserting that 'method has perhaps outgrown theory, and much work remains to interrelate time-budgets to society, economy, ecology, habitat and technology'. Have other disciplines come forward with more successful undertakings towards this more theory-based interrelation?

Let us take a closer look at anthropology's endeavours in this direction. Contrary to the quantitative sociological tradition, earlier anthropological studies have generally relied more on qualitative descriptions, sketching the broad outlines of time allocation in the communities studied. A screening of various anthropological literature sources reveals that the goals of time allocation studies have largely been the same: to provide insights into the sexual divisions of labour within a given community, to point to the social limitations of labour time through the lens of culture and to measure labour efficiency and effectiveness. What we see is that the study of labour time has been an attractive feature of anthropological time use research and, in contrast to sociological endeavours, generally been more theory-led. Sahlins (1972), for example, made a fine effort in theory-building as he mobilised, however heterogeneous, quantitatively available time use data and came up with the view that labour time increases as societies become more complex. This argument is neatly captured in his classic text *Stone Age Economics*, where he most compellingly dedicates the first chapter to the 'original affluent society' of hunters and gatherers.

Other early economic anthropologists like Salisbury (1962) or Firth (1967) have taken the classic economic idea of scarcity and choice as their point of departure. What I found, however, was that debates were primarily concerned with the economic sector of society, leaving aside the other domains that are equally important for the functioning of a society.

A different approach to the study of work comes from the ecological anthropology field. Rappaport's (1968) detailed monograph *Pigs for the Ancestors,* whilst striving to document the interdependence of cultural phenomena and biophysical variables, provides interesting data on energy expended in labour processes through the application of time-and-motion studies. While his analysis has been criticised on the grounds of the techniques used (see Ellen 1982: 115), his data has provided an impetus for similar kinds of research (see, for example, Pimentel and Pimentel 1979). More recent anthropological publications on time use among horticultural societies include Johnson's (1975, 2003) account on the Matsigenka of Peru and Descola's (1996) study of the Achuar ethnic group in the Ecuadorian Amazon, both of which have substantially contributed to the further refinement of time use methods.

This colourful anthropological assortment of time allocation studies also has its drawbacks. In his publication *Time Resources, Society and Ecology*, Carlstein (1982: 333) criticised the scant consensus on how to collect, analyse and present time use data. His second critique entails the missing link between time use records and demographic data. Without demographic data, he argues, time use figures are of no real use. The relevance of his critique, to my mind, still holds true today and should be widened to include also other social science disciplines. The trouble with time use studies is also the fact that, in addition to its objective characteristics (e.g., time spent engaged in a certain activity) time also has a subjective dimension. To this end, the limited agreement and consensus on concepts and analysis among the range of disciplines applying time use data research heightens the risk of producing misleading data. The classification of work and leisure, for example, reminds us of the subjective dimension of time use, as the categories 'work' and 'leisure' may be interpreted differently by diverse researchers. As to Carlstein's second critique on the missing link between population and time use, I also found that research papers on time use are generally only moderately explicit about the sex/age groups of the population studied. Lacking the precise terminology of data generation,[1] constructing scientifically valid comparative data becomes quite problematic. Finally, during my research into existing time studies, no conceptual tools that capture the changing dynamics of time use were encountered. 'Dynamic', in this respect, should be rather taken as a holistic approach. Rather than dividing time into

[1]Even Allen Johnson (1975, 2003), a strong advocate of time use studies in anthropology and a pioneer in terms of activity coding and classification among non-market societies, omits some of the terminology and demographic data of the Matsigenka population studied by him. In Carlstein's (1982: 365) words: 'Johnson's data is excellent in terms of time allocation for the relevant age-sex categories, but a minor flaw in his tables is that the exact number of individuals in each category was not stated'.

separate units, wouldn't an analysis be more rewarding if we understood how these units are kept together, and how changes in time use for one activity have a (dynamic) impact on the time budget for others?

After this rather long detour into existing time use research, here is what I am interested in: to study the dynamic effects of human time use on the social metabolism of Campo Bello. Or, more precisely, to gain insights into the social and environmental costs the community's time use profile entails. To this end, I have adopted some of the existing coding and classification methods and adjusted them to my needs. The overall theoretical frame, however, is provided by the MEFA framework (see Fischer-Kowalski 2003, 2007). In this framework, human time is seen as a limited key resource, given to princes and paupers in equal measure. The quantity of time available in a society depends on the number of people and their reproduction rate. In contrast to other sociological time use, human time is regarded as an element of social functioning rather than a personal resource of individuals. Human time use is distinguished according to different functional subsystems; this is based on the notion that each subsystem requires time for its functioning and is self-referential in the sense that it can reproduce itself only. At the same time though, changes in time use in one subsystem have an impact on the time budget available for the other subsystems, since the total time resources cannot exceed 24 h per day and person.

These changes in time use can take place via two sets of strategies: expansion/shrinkage, on the one hand, or intensification/de-intensification, on the other hand. Each set of strategies entails pressures on humans and the environment. Let us first examine the former set of strategies. Expanding time use on certain activities means reducing time use on others, e.g., expanding labour time may mean reducing sleep. In the long term, this may lead to social costs, such as conflicts or illness, which are provoked by chronic lack of sleep. In pre-industrial societies, we also see a direct link between the expansion of labour time and environmental costs, since all ways of interacting with nature are mediated through labour. The more a society colonizes its environment, the more metabolic returns it can expect, but the more labour needs to be invested in order to maintain the environment in the desired colonized state. Historic hunters and gatherers, for example, lacked the incentive to increase time investments in labour, as they, for the most part, depended on natural reproduction cycles. This direct link between the colonization of natural systems and the investment of human labour changed, however, with the onset of industrialization when labour became conceived as separate from the worker. With the tapping into fossil fuel resources and the industrialisation of agricultural practices, the extent of labour time no longer had a direct bearing on the environment. Here we enter the second set of strategies: intensification/de-intensification. Throughout history, time-saving assets have been invented to increase the productivity of time. Today, we travel from London to New York in just 5 h, while our grandparents would have had to calculate a fortnight for the same distance. The drawback with these ever more modern technologies, however, is on the input side. The more efficient and faster the technologies, the higher and more energy-intensive inputs usually are; a strategy that, in the shorter and longer run, works much to the detriment of the natural environment.

Admittedly, my aims are high. Measuring quantitative process dynamics in time use is hardly possible in the absence of longitudinal data sets and/or formal models. Adding to this, the existing framework undoubtedly needs some further refinement and empirical testing. What the concept can offer though, is a system level perspective on time investments into all functional subsystems segregated by age and gender. This provides us with is the opportunity to observe how time use changes in one subsystem affect other functions, as well as have an impact on the system as a whole. The use of these functional subsystems (largely) avoids the ambiguity of coding, as certain activities like sleeping or eating, which are physiologically necessary for basic personal reproduction, can unequivocally be categorised within the person system. What this framework also provides is a general activity and time allocation analysis, embracing all kinds of categorised activities throughout a 24-h time span. The advantage of such a complete record of the activities performed by different categories has the advantage that 'comparative theorists are then free to define variables in accordance with their theoretical aims rather than having to accept the incommensurable figures each idiosyncratic fieldworker may choose to publish' (Johnson 1975: 307). This may be particularly useful when analysing work-leisure dichotomies that have caused tremendous ambiguity among the time use research community. Finally, the systemic link between demographic reproduction, socially available (labour) time and actual time use grants some insights into current system dynamics while pointing towards possible and impossible futures.

5.1.2 Explaining the Functional Subsystems

Every social system comprises functional subsystems, all of which need time for their functioning. For the purpose of this study, we shall distinguish between the person system, the household system, the economic system and the community system. The person system functionally serves personal reproduction and includes all those activities that are not subject to the division of labour. On the one hand, the person system holds all the functions that are physiologically necessary for a person's self-reproduction, such as sleeping and eating. These activities can neither be delegated to other members of the society nor 'outsourced' to specialists and are largely horizontally distributed in a population's time-budget. Carlstein (1982: 313) entertainingly remarks that 'if it had been possible in human history to delegate sleep to other members of society, it would be more than likely that a special class of people would sleep on a full time basis so that the powerful could seek pleasure and practice various pursuits on a round the clock scale'. Yet there is no need to linger over this witty thought, as the scope for delegation of these activities is, so to say, non-existent. Apart from these basic functions for personal reproduction, the person system equally constitutes the functions for extended reproduction, such as studying, leisure activities or simply day-dreaming.

The second functional subsystem is the household system. Again, it comprises two functions, both of which are indispensable for the overall running of the domestic unit. They entail the intrinsic functions for basic day-to-day reproduction (e.g., cooking or shopping), on the one hand, as well as the functions that ensure the long-term maintenance of the household (e.g., care for dependents or house building). The household system is frequently organised as an exchange of unpaid labour according to the intrinsic social norms prevalent in the system under discussion. Reaching beyond the confines of the household, we find the economic system; it is the probably the least clear-cut subsystem when applied to subsistence societies. In general terms, the economic system hosts most of the activities that shape the social metabolism of a society. Many non-market societies, however, do not recognise the concept of 'work' as an activity separate from other types of social activity (contrary to the more unambiguous division between wage labour and household labour in modern societies). Equally difficult is finding a precise dividing line between purely economic functions, such as 'food provision', for example, as opposed to a household's 'food preparation'. For the purpose of my study, I have included both preparatory tasks for economic activities and transaction (e.g., manufacture of hunting and fishing gear, handicraft for subsequent sales), as well as directly productive work (e.g., agriculture, fishing, hunting) within the economic system. The economic system also provides the opportunity to distinguish between labour for subsistence and labour for the market. Finally, the community system is the reference system for activities contributing to the reproduction of reciprocal relationships, social cohesion, culture and religion. In non-industrial societies, this system may be regarded as a predecessor of several other, more specialised systems such as politics, religion or the judicial system (see Fischer-Kowalski et al. forthcoming).

5.2 The Collection, Coding and Classification Process

When it came to choosing data collection methods, I felt more at home with anthropological field methods rather than those common in sociology. The use of time diaries, for example, would have been highly inappropriate due to the fact that many people in Campo Bello do not have the necessary literacy levels required for this method. I opted for participant observation as my main method for data collection in the field, in order to gain entry into the 'backstage' life (see Goffman 1959) of the community. Only by doing so, would I be able to learn the meanings of different behaviours, intentions, events and situations.

Empirical time use research was carried out on various days between September 2004 and February 2005, followed by additional observations in April and May 2006. The time lag was important insofar as to capture seasonal variations, however slight. In order to study the differential time allocation patterns in Campo Bello, I originally selected the individuals according to demographic characteristics (various age/sex categories), migratory backgrounds and household composition. Notwithstanding, as two individuals originally selected decided not to partake in the time allocation sample, other individuals from befriended households, who had no objections to my rather intrusive 'shadowing', offered their participation. In the end, the final sample comprised 18 individuals; six married men (age ranges between 21 and 58), six married women (age ranges between 18 and 54), three girls and three boys (age ranges between 6 and 13) who were each observed for a 14-h daytime period (6 a.m. to 8 p.m.). From these individuals, I sampled three women, two men, one boy and one girl also the following year. Average activity profiles were thus calculated on the basis of a total of 25 person days put in by 12 men days (eight men and four boys) and 13 women days (nine women and four girls), spread over almost the entire annual cycle (except for June and July when no samples were taken). The male population sampled were aged 7, 9, 13, 21, 27, 33, 39, 45 and 58, whereas the females observed were six, nine, 12, 18, 26, 32, 36, 46 and 54 years of age. It should be noted that my samples do not include any 14-h observation of toddlers and small children below the age of 6. The main drawback was the fact that I could not observe a small child unaware. Children would all too often change their daily plans in order to show a visitor like me around the village and its environs. On a similar note, my sample population does not include any elderly residents above the age of 60, as the elderly village population was generally more apprehensive in participating in the time use study. Here is probably where the drawbacks of this research method become apparent, since direct observation, by definition, lacks a systemic sampling procedure.[2]

In an attempt to increase data validity and obtain more insights into daily time use patterns, I decided on additional spot sampling. In retrospect, it was indeed a good complement to full-day participant observation, since casual visits are a culturally embedded form of socialising in Campo Bello. Quite commonly people pay short visits to family and neighbours to catch up on the latest news or village gossip. 'The trick is', according to Bernard (2006: 426), 'to catch a glimpse of people in their natural activities before they see you coming on the scene and before they have a chance to modify their behaviour'. A non-random person was selected, who were visited four different times throughout the same daily round, taking discrete notes of what he/she was engaged in for a 15 min period each time. In so doing, a time allocation period of 1 h could be documented. This sample method was repeated for a total number of 14 active adults aged from 25 to 45 (seven men and seven women) and 14 children aged from 5 to 12 (seven boys and seven girls), from which an additional 14-h person-day could be derived. To sum up, the total of 112 spot check observations resulted in two full person-days which were consequently added to the pool of samples, thereby arriving at a total of 27 person-days. During the course of fieldwork, I found it extremely useful to resort to two different notebooks. The first I always carried with me; it was my official field book for primary data, containing all my field jottings, diagrams, interviews and time-space-maps. The second notebook served more as a personal memo book and contained any peculiar observations, quirky notes, 'off the record' contemplations, and field diary entries. Having both for different usages, the first and 'official' notebook was kept fairly orderly and uncluttered with extraneous field notes. In addition to direct observation and spot checks, my time use samples were complemented with household surveys. As people were asked about the main activities they engage in during the

[2] In order to enhance data variability, standard deviations were calculated for each of the four subsystem levels.

course of a week, this method proved particularly useful for getting a better idea of the duration and frequency of (especially) agricultural activities. A total of eleven interviews were conducted and cross-checked with other empirical data.

Moreover, to obtain data on cash flows entering the local system, interviews were carried out with all adults in a household. This was considered important for obtaining gender-disaggregated economic data. These surveys were conducted every 30 days in order to get a better understanding of seasonal trading activities. All interviews aimed to obtain information on cash incomes, on the one hand, as well as information on the type of goods purchased on the market, on the other hand. As to wage labour, records were made on the specific labour activity and the daily wage earned. Finally, for the study of labour time inputs for rice cultivation, a focus group was organised. The meeting was held in the local school building with 18 women and men constituting the group. The participants were asked about the number of work days required to do all the different tasks within the entire rice production process.

Every data collection process entails specific drawbacks and challenges, especially, as in my case, when resource capacities are rather limited. To begin with, I had to fall back on overall estimates at times, as the sample, as previously discussed, does not cover all age/sex categories, leaving out the male age groups from 16 to 20, 46 to 50 and 51 to 55. As with the female population, no females falling within the age groups 21–25 and 41–45 were directly observed. Adding to this, one may have noted that the total number of observed person-days represents a small sample only, accounting for just a little over 10% of the total population.[3] As to the particular field challenges, three are worth a closer mentioning: night time sampling, multiple and discontinuous behaviours, as well as, at certain instances, the distinction between economic food provision and a household's food preparation.

The first challenge was how to overcome the lack of night time sampling. Visiting others after dark meant not only engaging in perilous travel through the forest, but was simply seen as intrusive behaviour. Life though, does not stop when the sun goes down and the hours of darkness not directly observed could not be simply equated with 'sleeping'. Luckily though, two neighbouring households were situated close-by, whom I was able to observe from my dwelling. On many nights, therefore, I would simply sit out in the dark and lean against the wooden beams of my home, where I had found a perfect place to unwind from the day's events. These night by night observations provided some estimates on how the hours of darkness are generally spent. Rather than simply coding the nocturnal cycle as 'sleeping', time for hygiene, eating, resting and sleeping was allocated for men, whereas the women's activity profile comprised also childcare in addition to these activities.[4]

---

[3] There is no 'right' sample size engraved in stone, according to Bernard (2006: 166-167), however, as it depends on (1) the heterogeneity of the population or strata or clusters from which you chose the elements, (2) how many population subgroups you want to deal with, (3) the size of the subgroup and (4) how precise sample statistics have to be.

[4] For men, I took the following estimates per night (8 p.m.–6.a.m.): 30 min for hygiene, 60 min for idleness, 30 min for eating and 480 min for sleeping. For women, I took the following estimates per night (8 p.m.–6 p.m.): 15 min for hygiene, 120 min for child care, 15 min for eating and 450 min for sleeping.

The second challenge was the typological problem of multiple and discontinuous behaviours. Generating data for time use analysis is necessarily arbitrary since people engage in several overlapping activities. This was particularly noticeable in the case of household tasks, when women engage in general household chores at the same time, such as breastfeeding and tending the fire. Or what about simply eating and listening to the radio? Confronted with this problem plus the practical unfeasibility of timing the sequence of some activities (e.g., hunting and gathering), I was somewhat grateful for White's (1976) advice, who was facing the same difficulty during the study of a village in Java. He simply accounted half of the period to each activity. On occasion though, when the dominant activity was obvious, at least in my subjective perception, I only accounted for the sequence of the primary activity. With regard to the often discontinuous operations that constitute household tasks, for example, I generally had to fall back on overall estimates. A common sight was a woman's engagement in delousing her child, when she interrupted her activity with a sudden rise on her feet to stir the soup pot simmering on the open fire, then sat down again on the floor mat to resume delousing. Commonly, this scene would then be repeated various times.

The final and probably most taxing challenge was the functional distinction between economic food provision and a household's food preparation. When a manioc tuber is peeled directly in the field where it is uprooted, does this count as food preparation time or as part of a woman's agricultural activities due to its in situ performance? Or what about manufacturing a mortar and a pestle for husking rice, part of which gets sold at a later stage? It seems that the more detail we go into, the more blurred the picture becomes. In need of a practical solution, this is what I decided on. Regardless of the spatial performance of tasks, a woman's peeling of tubers in the field was still accounted for as food preparation as it is the first step in manioc beer production. As to the manufacture of household items for direct food production (such as a mortar, for example), the time invested was subsumed under the household system (despite eventual partial selling of rice). Only the manufacture of directly productive hunting or fishing equipment, as well as handicraft clearly destined for later selling, was subsumed within the economic system.

As concerns the coding and classification of all activities, Johnson's (1975 in Carlstein 1982: 365–368) time allocation study of a small indigenous community in the Peruvian Amazon was instructive for my own undertaking. Taking his activity code as a base, I adapted it to my needs, restructuring, adding or eliminating some activities not considered of too much relevance for my own purpose. In a further step, every activity was assigned to one of the four functional subsystems. The single activities were classified as follows: (1) person system: Sleeping (SL), eating (ET), hygiene (HY), rest and idleness (ID), leisure activities (LE), studying and education (SC); (2) household system: care for dependents (CC), food preparation (FP), house building (HB), repair/maintenance work (MR), domestic chores (D)[5]; (3) economic system: agriculture (AC), hunting (H), fishing (F), gathering (G), trading (TD),

[5]The category 'domestic chores' entails firewood collection, the fetching of water, washing laundry and kitchen utensils, and general domestic chores.

wage work (W), housegarden (HG), handicraft (MF), animal husbandry (AN); (4) community system: public sports and games (PL), visiting friends and relatives (VS), ceremonies and festivals (RI) and community and political participation (PO). Concerning the latter, I did, however, account for the time invested in communal work separately in order to gain insights into the importance of communal gatherings as opposed to family settings. Though for reasons of simplicity, communal work was subsumed under community and political participation.

5.3 The Qualitative Order

5.3.1 Introducing the Daily Round

The people of Campo Bello wake up at first light around 5.30 a.m. and every other morning, women get firewood from nearby fields or the adjacent forest. Around the same time, water is brought back from the source, a task that can be liberally delegated to children. When the water starts boiling over the crackling sound of the fire, the preparation of breakfast has began; women husk the rice, cook some dried meat or simply grate and boil plantains. It was interesting to note that people were very reluctant to call a meal without meat a 'proper' meal. Not uncommonly, men go out on a bird hunt in the early morning mist in order to satisfy their family's demands for meat. Just like in our societies, breakfast is typically accompanied by the sound of the radio. Most mornings, people listen to the religious radio programme launched by the New Tribes missionaries starting at 7 a.m. Held in the Tsimane' language, these religious programmes are quite popular and mark important family gatherings throughout the day; at 7 a.m., at 1 p.m. and at 7 p.m. This time of the morning is quite a leisurely period as individuals start preparing for the day. With the women tending the fire, older children get ready for school. Shortly before 8 o'clock, children are reminded by the sound of the school horn to make their way to school, a journey that normally takes between 10 and 30 min.

By that time, a man gets ready for his working day. He snatches his woven bag filled with utilities for the day ahead; a hunting trip or a working day in the field. Agricultural labour, hunting, or other manual work fills out most of the morning hours. Since most fields are located near the family dwellings, travelling time is quite limited. Other days not dominated by agriculture may be dedicated to house building or the occasional communal chore. Tsimane' men are vigorous labourers and most productive in the morning hours, when they work hard and fast without enjoying too many breaks. By early afternoon, with the heat becoming unbearable, work is usually left for the next day and leisure time begins. Afternoons are repeatedly taken up by bathing, resting or simply socialising with family and neighbouring men. A woman's work day, on the other hand, tends to be parcelled out in a whole host of interrupted, repetitive chores like tending the fire, washing dishes and clothes, carrying water or taking care of dependents. Within the house, one really gets the impression that a woman follows a schedule that is nearly identical day in day out. Time and again, a woman's late morning hours are used for agricultural chores such as the planting of manioc or weeding. These activities are usually abandoned when the midday heat becomes too strong. Food preparation begins around this time. Husking rice takes up most of a woman's time during food preparation and is accomplished by using a heavy wooden plank with a pestle that is operated in a see-saw fashion. The time after lunch is dedicated to some rest in the shade. It is a time of leisure, often accompanied by the sound of Bolivian folk music buzzing from a transistor radio that is always close at hand.

Afternoons are usually devoted to manufacturing activities. Men are often seen manufacturing a basket or sharpening their darts, while women dedicate a substantial amount of time to weaving or the fabrication of floor mats. On most days around 5 p.m., boys and girls gather to play football or volleyball. Attracted by the social event, young mothers and older women get together in front of the football pitch, exchanging the latest gossip. As dusk approaches, families make their way back home and women briskly start preparing the fireplace. Frequently, the tone of the radio emitting the daily 7 p.m. ritual is mingled with quiet conversations recounting the events of the day. Family members gather near the fire, the dim light of the night lamp reflecting on the peoples' tired faces. To get ready for the night, mosquito nets are fixed, floor mats rolled out and piles of clothes spread out. By 9 o'clock, families are usually settled for the night.

As a general observation, daily activity patterns change little throughout the different seasons. On rainy days, however, activity levels drop sharply as work becomes almost exclusively bound to the house. While the fireplace presents an important place around which people gather to warm themselves, men and women often use these home-bound moments to engage in manufacturing activities. It seems that people get more rest during the rainy season in January and February, as evening leisure is often constrained by the presence of hoards of mosquitoes. After heavy rains, mosquito nets frequently offer the only safe refuge. While some activities change little throughout the year, labour investment in agriculture does, however, undergo seasonal cycles.[6]

5.4 The Quantitative Order

5.4.1 Socially Disposable Labour Time

Anthropologists have found strong evidence that time investments in labour in many indigenous societies are socially organised (see, for example, Descola 1996; Netting 1993). Or to put it differently, capabilities and capacities are not purely a function of biologically based variables, such as age and gender, but culturally transmitted and

[6] For a detailed description of the agricultural cycle please refer to Chapter 4.

socially ascribed. A strongly pronounced sexual division of labour pattern, for instance, would be a fine example of such social ascriptions. At the same time, the availability of the overall labour time in a community depends on the number of people living in the system and their demographic reproduction rates. Carlstein (1982: 393) asserts that life-cycle events such as births, deaths, migrations and household compositional changes over time are not simply biological 'accidents' affecting society, but very much endogenous process dynamics. I am interested to explore these aspects in the following section. Looking at the socially disposable labour times by age/sex group aims to focus on the specific capacity and capability constraints faced by each age/sex group within a certain life-cycle period.

To determine the disposable labour time available in Campo Bello, the average duration of each activity was allocated according to the four functional subsystems (see Table 5.2.). By so doing, a time aggregate for each individual subsystem was obtained by sex. These times were subsequently allocated among the age/sex categories to arrive at a 24-h period for each age group. The time available for the reproduction of the economic system was termed the 'socially disposable labour time'. In order to arrive at the disposable annual labour time of the different age groups within the community, the disposable daily labour time was multiplied by the amount of people within the age range and the 'work' days of the year. Concerning the number of 'work' days, the number estimated was 300, accounting for festive days, days away from the community, and occasional sick days (Table 5.1; Figs. 5.1 and 5.2).

Looking at the demographic profile of Campo Bello, we have learnt elsewhere in this book about the strikingly high number of infants between the ages 0–5. Despite the fact that no children falling within this age group were actually sampled, it is evident that they do not have the physical capacities to invest time in productive labour activities. Having said this, however, children do start to contribute their labour within the household from an early age. Befitting a traditional society, the transitional process for children in Campo Bello, from completely dependent recipients of parental time to more active economic producers, starts at a relatively early age, some time around the age of 4 or 5. Children start engaging in 'enabling labour'[7] activities, performing physically undemanding tasks like carrying bowls of water, fetching items and shooing chickens.

Moving up the age scales, labour time resources for 6–10 year olds become more pronounced. Children, albeit in a playful manner, have become active in gathering forest products, fishing and the occasional agricultural chore. As for boys of this age group, the total disposable labour time per day was determined to be 50.6 h/day (15,180 h/annum). Due to the high number of boys falling within this age category, it is in fact the age group with the highest potential labour time at their disposal. Girls are equally active, especially in gathering forest fruits or fishing. On the system level, however, girls invest less time in the economic system due to the

[7]This term is borrowed from Cain (1980) as it frees other household members to engage in activities that are directly productive.

Table 5.1 Potential labour time supply according to different age groups, Campo Bello, 2004, 2006

	Female				Male			
	Nr.	Disposable labour time person (h/ day)	Disposable labour time community (h/day)	Disposable labour time community (h/annum)	Nr.	Disposable labour time person (h/ day)	Disposable labour time community (h/day)	Disposable labour time community (h/annum)
0–5	41	0.0	0.0	0.0	35	0.0	0.0	0.0
6–10	17	2.4	40.8	12,240	22	2.3	50.6	15,180
11–15	10	2.8	28.0	8,400	15	2.8	42.0	12,600
16–20	7	3.3	23.1	6,930	6	4.4	26.4	7,920
21–25	8	2.9	23.2	6,960	7	5.6	39.2	11,760
26–30	7	2.9	20.3	6,090	5	6.2	31.0	9,300
31–35	2	3.5	7.0	2,100	3	6.4	19.2	5,760
36–40	3	3.7	11.1	3,330	6	6.5	39.0	11,700
41–45	4	3.8	15.2	4,560	4	6.5	26.0	7,800
46–50	5	3.5	17.5	5,250	1	6.5	6.5	1,950
51–55	2	3.5	7.0	2,100	3	5.8	17.4	5,220
56–60	0	3.5	0.0	0.0	6	4.4	26.4	7,920
>60	6	2.3	13.8	4,140	6	2.5	15.0	4,500

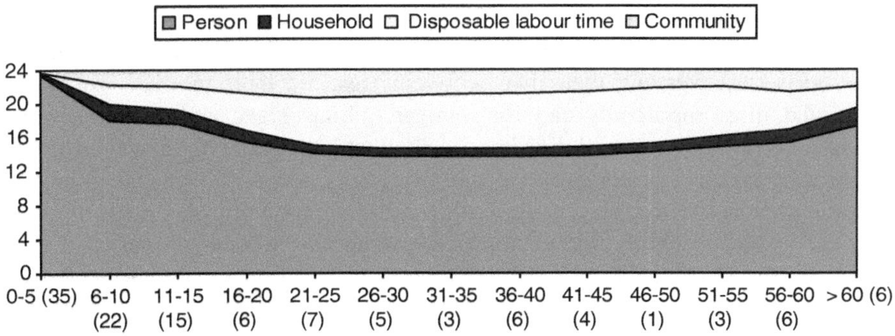

Fig. 5.1 Male time use by age group, Campo Bello, 2004, 2006

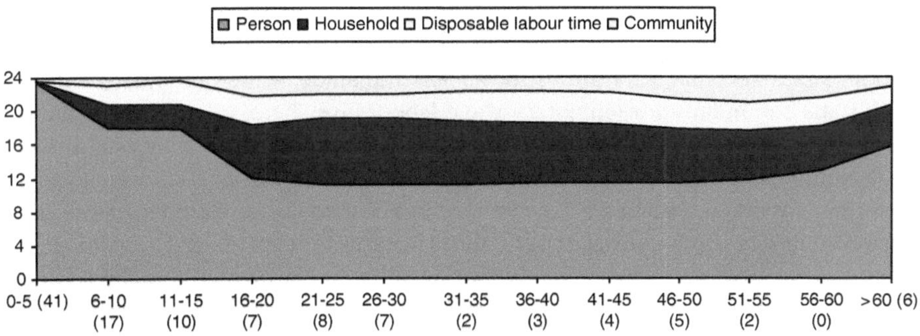

Fig. 5.2 Female time use by age group, Campo Bello, 2004, 2006

lower number of individuals belonging to this age group (12,240 h/annum). To a large extent, physical capability constraints prevent the boys and girls from engaging in more labour-intensive activities. Girls, for example, may help with the production of manioc beer and other light food preparation tasks. Having said this, however, a minimum weight is required before a girl can physically operate the cumbersome pestle for husking rice. Apart from physical restraints, educational demands on the children also limit their disposable labour time. In 2004, the local school system comprised five grades and, except for two families, all children between the ages of 6–14 (except for those who had already married or had a baby) regularly went to school. School classes consume 4 h every day and additional time for homework in the afternoon.

For both sexes, this age range represents a crucial socialisation period. It is a period of playful learning when one still enjoys the light-heartedness of being a child, engaging in adventurous gathering trips to the nearby forest. At the same time, parental guidance gradually prepares them for their respective future roles. Boys generally aid with house building and maintenance work, often collecting lashes and palm leaves in the environs of the house. Young boys accompany their fathers on hunting expeditions, join family fishing trips and engage in physically undemanding agricultural tasks such as the transport of plantains or fetching of seeds. Inside the house, girls start weaving their own woollen bags, aid their mothers in sewing clothes and, most importantly, care for younger siblings. Have said this, however, I have also noted that the sexual division of labour, at least in the initial stages of this socialisation process, is not imposed too stringently. Boys still engage freely in domestic chores like carrying water, while girls occasionally play with wooden bows and arrows or simply find great pleasure in carving wooden knives.

By the ages 11–15, children are turning into adolescents, not only physically but also socially. Round about the age of 12, a boy starts to cultivate his own agricultural fields, even if this seems to be more of an educational activity (see Piland 1991). In terms of labour provision, 15 boys and 10 girls were counted within this age group, accounting for 42 h/day (12,600 h/annum) and 28 h/day (8,400 h/annum) of disposable labour time, respectively. With physical capacity constraints diminishing, educational demands still present the most limiting factor for the disposal of more labour time. In 2004, the proportion of boys and girls from 11 to 15 still attending school classes accounted for almost 80%. The remainder were one older boy who had left the community for temporary wage labour elsewhere, two girls who had to look after their babies and one recently married couple who were busy establishing their own household. For the most part, individuals from 16 to 20 are engaged in preparing, directly or indirectly, for family life. Those male adolescents who remain in the community begin to work in agriculture in earnest, often tending several active fields at the same time. Building one's own dwelling in close vicinity to the parental home is also quite a common practice for some young male adults; I was told that such an endeavour would be a clear sign for an upcoming marriage. Other young males use the time that was previously taken up by studying for engaging in wage labour outside the community. For a few weeks at the time, they may become employed as farmhands or crew members on itinerant trading boats. If we look at the Tsimane' girls, the transition to married life is more abrupt. By the age of 16 or 17, most girls are married or have an infant. This is the reason why, despite their slightly higher numbers, girls can only contribute 23.1 h/day (6,930 h/annum) to the economic system. Adolescent boys, on the other hand, have 26.4 h/day (7,920 h/annum) of labour time at their disposal.

From the ages 21–25, the investment of male labour is to a large extent used up in outside wage labour. Some are pressured by former employers, as often the accumulation of debts from previous labour engagements forces them to continue with this kind of work. Others simply seize the opportunity of temporary wage labour to save some cash for their recently established households. In a Tsimane' man's life-cycle, this is probably the most flexible and 'freest' time of his life. In 2004, no women were observed to engage in wage labour in Campo Bello. Contrary to her male counterparts, a woman's time resources are largely used up by reproductive activities for the household system. With small children to take care of, a woman almost exclusively has to bear the time costs of child rearing. It is in fact one of the most time-consuming and difficult stages of a woman's life. If the husband is away on temporary labour, a woman of this age not only bears the reproductive work-load, but also assumes the productive burden to provide enough food for her infants. To facilitate this, I have observed a few young women, at least temporarily, move back to their parent's home.

The reproductive work-load for women starts to decrease steadily over the following years as older children gradually step in and take over some of the activities. One important way for children to contribute their time and effort is to tend younger siblings, thereby imposing fewer time demands on their mothers (see also Carlstein 1982: 396–98). When this is the case, mothers are freed to dedicate their time to more productive activities such as gathering or agriculture. Men, on the other hand, still dedicate some of their time resources to occasional wage labour activities; a large fraction of their working time, however, is used up by agricultural activities. In Campo Bello I found that male residents have a total of 39.2 h/day (11,760 h/annum) at their disposal for directly productive labour tasks, whereas female residents, albeit slightly higher in numbers, can only dedicate a total of 23.2 h/day (6,960 h/annum) to the economic sphere. The following years are the most productive for both sexes in terms of agricultural labour investments. This is due to the fact that older children are useful farmhands, and add considerably to the household's labour force. Interestingly, Piland (1991: 107) found that families with teenage daughters produce more rice than others, while families with teenage sons produce notably more plantains than families without adolescent sons.[8] Men and women often work simultaneously on various active fields; if we look at the area of cultivated fields and fallows we find that the amount of fallows increases with a Tsimane' farmer's age. This correlates with what we have learnt about the Tsimane' inheritance rules, namely that fallows stay within the same family and are only passed own within that family. Between the ages 31–35, male villagers are able to invest a total of 19.2 h/day (5,760 h/annum) in productive labour. This small amount is a result of the low number of individuals comprising this age group. The women of this age group, on the other hand, can dedicate only about half a man's time to the economic sphere. Adding to this, the low number of females generates a contribution of a mere seven labour hours per day (2,100 h/annum) for the whole system. An examination of the age groups between 36 and 40 reveals that local males can supply a total of 39 h/day (11,700 h/annum), whereas the contribution of female time resources, due to their lower numbers and labour time disposability, amount to only 11.1 h/day (3,330 h/annum).

By the age of 55 or older, agricultural work-loads have decreased (see fallow discussion above). Also, the wage labour market no longer holds an attraction for men of this age. In 2004, the only exception was Rafael, who, though in his late 50s, was a highly solicited farmhand at a cattle farm nearby. Due to their higher numbers, male villagers (56–60, older than 60) contribute a total of 41.4 h/day (12,420 h/annum) to the system's productive labour force, while the women's disposable labour time accounts for 13.8 h/day (4,140 h/annum) within the same age

[8] This finding, however, could not be confirmed as in my observation, teenage daughters would plant their own rice fields and so did sons.

group. As an interesting detail, at the time of research in 2004, no women fell within the age group 56–60. Above the age of 60, women engage a great deal of their time outside the household in gathering activities, whereas a man contributes with the manufacture of hunting and fishing devices, among other more directly productive chores. Generally though, the available labour time decreases around this age as both, men and women alike become more house-bound.

The above discussion reveals the importance of examining the demographic profile of the entire social system in detail, rather than looking only at an individual's capabilities. In theory, the potential contribution of labour may be highest in an age group that is free from physical, educational, and social constraints. In practice, however, the actual contribution of labour ultimately depends on the number of people constituting this age group. Having said this, however, it also works the other way round. If we look at the 6–10 year olds that comprise the second largest number after the toddlers, this age group would indeed be able to provide a substantial amount of labour if only aggregate numbers are scrutinised. Nonetheless, they are largely constrained by physical limitations as well as the educational demands that have become strongly embedded in the local culture in recent years. On the other hand, 35–50 year olds, although largely free from physical, educational and other social restraints, are equally limited in their potential contribution to the economic system. The reason for this lies in the low number of individuals living in the community. To sum up, this demographic analysis reveals the limited availability of people in their most productive years, coupled with high numbers of individuals in lower productive ranges, most of whom are economic dependents that add to a woman's time demands. Adding to this, a man's absence from the family due to outside wage labour, puts an additional burden on a young woman's time, as she must increase her own time investment in order to compensate for that of the absentee. Since not all of the work usually done by the husband can be taken over by the women (e.g., hunting), even if her workday expands, food or nutritional shortages for her and her family may be the result. This is where the extended family network steps in, to a certain extent providing a 'safety net' for these periods.

5.4.2 *Time Investments in All Functional Subsystems*

In this section, I want to expand the scope of analysis from a mere focus on disposable labour time to examine all functional subsystems, the person system, the household system, the economic system and the community system (Table 5.2). Here, the overall goal is to gain insights into the number of person-hours invested for the reproduction of each of these subsystems. The specific interests are twofold. First, what is the scope for increasing labour inputs into the economic system? To some extent, this aim is an extension of the above analysis, this time however, accounting for all the basic and extended activities inherent to a person's reproduction. The economic sphere, as we have come to know, provides a direct link to the system's social metabolism, as this system hosts most of the activities taking place at the society–nature interface (except for trading and wage labour). My second, though interrelated, interest lies in scrutinising the work allocation between both sexes. How is labour distributed among the sexes and what would the environmental and social impact be, if certain activities were expanded or intensified?

5.4.2.1 The Person System

As the person system is the only social system that can provide time, I have calculated a maximum annual time budget of 2,2023,560 h for Campo Bello. If we examine the time invested in the reproduction of the system, we see that 66% of all available time resources are taken up by this system. What seems strikingly high at first, becomes less surprising if we look at the individual activities that are hosted by this system. Some tasks require a certain minimum amount of time that needs to be invested by each member of the society and cannot be delegated or transferred to other categories or members of the social system. Starting with the functions for basic reproduction, I am going to examine the time investments into each of the activities that constitute the person system.

In Campo Bello, sleeping takes up a total of 1,995 h/day (728,175 h/annum), presenting by far the largest person-hour investments into the person system and overall. Disaggregating this number by sex, we see that women's time investments in sleeping account for 48% of total time inputs. Especially young women, at least for some years, tend to be deprived of some of their nocturnal sleep due to breast-feeding activities. The estimation of sleeping hours for children is slightly higher as compared to adults, since for their biological reproduction children of both sexes need more hours of sleep than adults (see Carlstein 1982). Eating amounts to 431 h/day (157,277 h/annum). It appears that time investments by both sexes are fairly uniformly distributed among the local population. The lower time investments by women does not manifest as relatively limited access to food, but has probably has to do with the fact that women tend to nibble more while away on a gathering trip; something not accounted for in the sampling process (Table 5.2).

Another intrinsic function of the basic reproduction of the person system is hygiene. A total of 101,800 h/annum are invested in grooming, washing and bathing. Grooming is an important activity due to the widespread preoccupation with head lice. Picking head lice is almost exclusively a female task and an activity that is engaged in at great length throughout the day. It starts with slapping the child's head and rooting around in the hair; once the lice are spotted, they are swiftly picked out and cracked between the nails. I have seen 7 or 8 year olds engage in delousing with the same patience as grown ups. After a long school morning, children like to indulge in extended river bathing, a playful activity that is accompanied by splashing and cries of laughter. Given the different rates of activity for men and women, men tend to have a more continuous break after a period of intensive and energy-expending morning labour in the field or forest. A bath in the Rio Maniqui provides for such compensation from hard work and men repeatedly invest more time in this activity than women. With the example of river bathing, the dividing

Table 5.2 Time investment in functional systems by age/sex category, Campo Bello, 2004, 2006

	Boys 6–15	Male adults 16–60	Male adults >60	Girls 6–15	Female adults 16–60	Female adults >60	Total h/day	Total h/annum	%	% Female		% Male	
Person system													
Sleeping	9.0	8.2	9.0	9.0	7.6	9.0	1,995	728,175		954	48%	1,041	52%
Eating	1.9	1.7	2.0	2.1	1.5	2.0	431	157,277		208	48%	223	52%
Hygiene	1.3	0.9	1.5	1.4	0.7	1.5	279	101,800		133	48%	146	52%
Rest and idleness	2.3	3.2	4.2	1.9	1.6	2.5	609	222,285		226	37%	383	63%
Leisure activities	0.3	0.0	0.6	0.4	0.2	0.6	86	31,390		47	55%	39	45%
Studying and education	3.0	0.3	0.0	3.1	0.0	0.0	248	90,583		116	47%	132	53%
Subtotal	17.9	14.4	17.3	14.9	11.6	15.6	3,648		66				
Household system													
Care for dependents	0.5	0.0	0.0	1.5	3.2	2.2	276	100,740		256	93%	20	7%
Food preparation	0.2	0.4	0.3	0.5	1.7	1.0	156	56,940		118	76%	38	24%
House building	0.0	0.4	0.5	0.0	0.2	0.0	40	14,600		6	15%	34	85%
Manufacture/maintenance	0.2	0.3	1.4	0.2	0.4	1.0	135	49,275		62	46%	73	54%
Domestic chores	0.9	0.0	0.0	0.7	1.3	0.8	139	50,735		105	76%	35	25%
Subtotal	1.8	1.1	2.2	2.9	6.8	5.0	746		13				
Economic system													
Agriculture/horticulture	0.8	2.6	1.3	0.2	2.0	0.8	299	109,135		115	38%	184	62%
Hunting	0.5	0.9	0.6	0.0	0.0	0.0	80	29,200		0	0%	80	100%
Fishing	0.4	0.5	0.3	0.7	0.2	0.3	90	32,850		44	49%	46	51%
Gathering	0.3	0.0	0.0	0.9	0.3	0.6	78	28,470		66	85%	12	15%
Trading	0.0	0.6	0.1	0.3	0.3	0.0	49	17,885		23	47%	26	53%
Wage work	0.0	0.5	0.1	0.0	0.0	0.0	24	8,760		0	0%	24	100%
Housegarden	0.0	0.1	0.0	0.0	0.3	0.2	22	8,030		16	73%	6	25%
Manufacture/handicraft	0.6	0.5	0.2	0.4	0.3	0.3	88	32,120		38	43%	50	57%
Animal husbandry	0.1	0.0	0.0	0.2	0.0	0.1	15	5,475		13	87%	2	13%
Subtotal	2.6	5.8	2.5	2.6	3.5	2.3	745		13				

(continued)

Table 5.2 (continued)

	Boys 6–15	Male adults 16–60	Male adults >60	Girls 6–15	Female adults 16–60	Female adults >60	Total h/day	Total h/annum	%		% Female		% Male
Community system													
Public sports and games	1.5	0.2	0.0	0.5	0.0	0.0	89	32,485		19	21%	67	75%
Visiting	0.3	1.7	1.5	0.2	1.5	1.0	234	85,410		97	41%	137	59%
Ceremonies and festivals	0.0	0.5	0.3	0.0	0.3	0.0	43	15,695		11	26%	32	74%
Community and politics	0.0	0.2	0.3	0.0	0.3	0.2	39	14,235		19	49%	20	51%
Subtotal	1.8	2.6	2.1	0.7	2.1	1.2	405		7				
Total	24	24	24	24	24	24	5.544						

line between 'hygiene' and bodily regeneration, 'rest and idleness' becomes somewhat blurred. Yet to make the functional division more clear-cut, it was decided to include only those periods of inactivity in the category 'rest and idleness', totalling 609 h/day (222,285 h/annum). What is striking is that male community members, children, adult men and men above 60, invest more hours, in fact almost two-thirds of all person-hours, in this reproductive activity. One reason, obviously, has to do with the nature of a man's daily chores. After periods of intense morning activity, a man can spend the remainder of the day in a restful pose. A woman's day, on the other hand, is more varied and contains less time and energy-consuming tasks. At the same time, my impression was that engaging in idleness seems much more socially acceptable for men than for women; a pattern that starts from an early age and, as the data confirms, remains true in old age. Leisure accounts for the last function that falls under basic personal reproduction and includes all leisurely activities such as playing with children and animals. Admittedly, the fact that merely 86 h/day (31,390 h/annum) are invested in this activity may reflect its somewhat vague character in traditional societies, where the work/leisure dichotomy is less clear-cut than in industrial societies. Also with regard to my case study, playing with children was subsumed under child care, if it was perceived as a secondary activity of breast feeding, for example. Hence, only those instances when playing was clearly observed as a distinct activity were subsumed under 'leisure'.

It is interesting indeed, that more than 90% of all person-hours invested in the functioning of the person system are invested in 'basic' reproduction. The remainder is invested in what Fischer-Kowalski (2007) calls the activities for 'extended' personal reproduction. This includes studying and education. Educational demands are fairly evenly distributed amongst boys and girls and, in 2004, almost all children regularly went to school. Parents generally support the idea of children going to school and attend school meetings on the progress of their children. Classes run from 8 a.m. until 12 p.m., with a half-hour break in between. Some time is also invested in doing homework, while girls generally seemed more diligent with these tasks than boys (an observation that was also confirmed by the teacher). In 2004, no adult women engaged in educational activities, whereas some of the younger adult men attended regular vocational training classes funded by a local NGO in the neighbouring community of San Antonio. At the time of research, they took place at regular intervals of about once a month. To round up the person system, we may note that men and women invest fairly the same amount of their time resources in the reproduction of this system, accounting for 53% and 47%, respectively.

5.4.2.2 The Household System

The household system constitutes the second functional system and, in terms of person-hours invested, accounts for merely 13%. What is similar to the person system is the fact that time needs to be invested to ensure the basic functioning of the household. What differs from the person system, however, is the fact that certain labour tasks can be shared or, in fact, completely delegated to others. Not all people within the household system therefore require the same time and energy inputs. Delegation dynamics are intimately linked to the socially embedded allocation of labour in a given society. A sexual division of labour inside the household is also inherent in Campo Bello. Having said this, however, I did observe a certain degree of flexibility on task gendering. This, I gather, is especially the case when opportunities to do otherwise are limited. Enrique, for instance, took over his wife's tasks of cooking and housekeeping when she was treated for tuberculosis in the regional hospital. Although this role conversion seemed quite unusual at first, he was highly respected for assuming the reproductive tasks that would have otherwise been done by his wife. Time and energy-wise, women are primarily responsible for the day-to-day running of the household, providing a fair amount of 'enabling' labour so as to free other household members to engage in more productive work. In fact, 60% of all time invested into the reproduction of the household system is provided by women.

Care for dependents takes up most of the time required for the reproduction of the household system and is primarily invested by young mothers with babies and infants (more than 90% of all child care is done by females). When children grow older, things start to change. As a general rule, once they have passed the first 3 years of their lives, they can be looked after by any older member of the family. From the data we see, that young children and elderly women can make a considerable contribution to household activities and, in so doing, release the mother to invest her time in more directly productive work. As an interesting observation, young boys also readily accept the tending of smaller siblings; a task that gradually diminishes as the socialisation process advances. Caring also includes a woman's task of attending the sick. Children particularly suffer from colds or respiratory infections at certain times of the year, while cuts and wounds are common throughout the entire year. Interestingly though, when people are really ill, it seems that the rest of the household goes on as usual, somewhat leaving the patient to his or her own devices. As Johnson (2003: 89) denotes, the distant behaviour towards sick kin may reflect an ambivalence arising from fear that the sick person has brought spiritual danger into the household.

Food production is mostly a woman's domain too, accounting for 76% of all female time investments. Except for the butchering of large animals and salting of game meat, men's contribution to food production is at best indirect. This has to do with the fact that a man's task is the manufacture of cooking utensils only, ranging from wooden spoons, grind stones, beer vats, or a pestle and mortar for rice husking. Once the meat is butchered, women take over to skin monkeys, pluck fowls or simply remove the husks and skins from fruits and vegetables. Foods are boiled in a large aluminium pot, mostly without much seasoning except for salt. The empirical study also reveals that sons gradually grow away from their mother's sphere of influence, while girls begin to make significant contributions to the woman's workload inside the house, learning the essentials of food and beer production. House building and maintenance, on the other hand, is a man's domain that takes up about 14,600 h/annum. Within this male sphere, a woman's participation is mostly restricted to gathering construction materials like palm leaves and balsa wood. Boys of 16 or older already start to build their own dwellings, often with the help of other male adults within the family network. It was interesting to observe that villagers are inclined to build new dwellings rather than investing too much time in general maintenance activities. As dwellings are built entirely from traditional biomass, a building's lifetime is restricted to about 5 years.

The manufacture and maintenance of household artefacts are continuous processes and 135 h/day (49,275 h/annum) are invested into the production of the essentials required for daily subsistence. Time used in manufacturing and maintenance work is slightly higher for men, accounting for 54% of male time investments. In light of sexual labour division, men and women tend to produce different commodities. If a man does not know how to make a given item of men's manufacture, his wife would have to obtain it from another man. The possession of these skills is therefore a matter of vital importance. We see that older men invest quite some time in fabricating household items, accounting for 1.4 h/day. Angel, for example, enjoyed a good reputation for fabricating different-sized *paasí*, a kind of tightly-woven sieve that is used for separating the manioc or plantain fibre from the beery liquid. The manufacture of toys also falls under this heading. Men make a variety of wood carvings, from small balsawood canoes to all sorts of wheels; or, in the absence of a proper football, creative minds swiftly make balls out of plantain leaves or engineer a spinner out of a pumpkin fruit. Fans are fabricated by both sexes, while the manufacture of floor mats and storage bags falls within a woman's realm. The weaving of woollen bags, however, constitutes by far the most time-consuming activity for a woman. In addition, domestic chores like sweeping the patio, washing clothes and dishes, or general cleaning also require a woman's time and energy. Having said this, however, the contribution of boys is indeed striking, constituting 0.9 h/day. Most of their labour goes into the transport of water and the collection of firewood. This empirical analysis shows the far less stringent sexual allocation of labour with children. As he grows older, a boy is no longer required to do these domestic tasks; a finding that is also reflected empirically. A woman's domestic work-load can be aided by an elderly woman supporting these chores. Mending her children's and grandchildren's clothes, for instance, was a task that Delmira engaged in every other afternoon, seemingly enjoying this contribution to her family.

5.4.2.3 The Economic System

The economic system represents the third functional subsystem and, just like the household system, accounts for 13% of all person-hours invested. When examining this system, it makes sense to differentiate between labour for subsistence and labour for the market; this will be discussed briefly towards the end of this section.

Agriculture is indeed the most time demanding economic activity, accounting for 45% of all time spent in the economic system. Labour requirements depend on the natural agricultural cycle of the cultigens and reach their peak with the rice harvest between March and May each year. A total of 109,135 h/annum are invested by both sexes, whereas men's labour time inputs account for 62%. If we look at the gendered agricultural labour allocation, the broad outline is as follows: Tsimane' men tend to plant a few crops in large amounts (i.e., rice and plantains), while women are more likely to plant various cultigens in smaller amounts, thus contributing to nutritional diversity and taste (e.g., manioc, peanuts, water melons, avocado). Whereas women are mostly in charge of planting and harvesting manioc, it is their husbands' responsibility to open new fields. In most cases, general field preparation is carried out by men. This is especially true for opening a clearing from primary forest, when work expenditure is extremely heavy. When done traditionally with a dibble stick, the planting of seed crops may involve a married couple and some of their older children. In recent years, however, manual rice seeders have cut the work-load tremendously. They are widely loaned among families and usually employed by men only. Once passed the planting stage, sexual labour division becomes less pronounced and weeding and harvesting are often done by both sexes. If we examine the time spent in agriculture, we arrive at an average of 2.6 h/day for men and 2.0 h/day for women. Compared to other swidden horticulturalists in the tropics, the indigenous Matsigenka from the Peruvian Amazon, for instance, where daily agricultural work amounts to 2.5 h/day for men and 0.9 h/day for women (see Johnson 2003), we find a woman's contribution to agricultural labour to be comparatively high. And still, if we compare these time investments to more permanent crop farming, the Tsimane', as swidden cultivators, do not need to invest time and effort in indirect labour, such as the transplantation of rice seedlings or the collection and distribution of manure. Likewise, the Tsimane' do not remove unburned tree trunks from fields and crop seeds are simply dropped into holes without turning or breaking the ash covered soil immediately afterwards.

Since agriculture is tied to seasonal factors such as rainfall and temperature and to natural cycles of reproduction and growth in plants, labour cannot simply be increased by lengthening the working day (see Netting 1993: 118). Cutting timber, for example, needs proper timing, in order to guarantee adequate drying. Likewise, if dense forest is slashed too early, weeds grow rapidly, and a bad burn may leave emergent undergrowth and pests to compete with the crops. Hence, simply expanding work days during peak periods would not be an option. Instead, the local families have other options to fall back on. First, they can pool the labour resources they have available within their own nuclear and extended households. This is eased by the less stringent labour division when it comes to rice production and all its labour demands throughout the cropping cycle. Another strategy to even out agricultural peak periods is to plant various crop varieties in order to space the cultivation, weeding and, above all, harvest periods.

Only male residents invest time in hunting, accounting for a total of 29,200 h/annum. From an early age, boys are occasionally taken on hunting trips and introduced to the secrets of hunting by listening attentively to the interminable hunting tales. Sometimes, fathers fabricate small toy bows and arrows for their sons to practice on small living targets like birds, frogs or beetles. Once past the age of 60, men may have lowered their work-load in agriculture, but the same does not hold for hunting and elderly men still bring home a good game catch every so often. Gathering, on the other hand, is an activity taken up more frequently by women; females, in fact, account for 85% of all person-hours invested in this activity. Little girls, for example, enjoy gathering a variety of seeds for the production of necklaces and other pieces of 'forest jewellery'. Little boys, on the other hand, are active in honey or fruit collection. According to Table 5.2, no time is invested in gathering by adult men and men over the age of 60. This, however, has more to do with a methodological weakness, since male gathering activities are almost exclusively combined with hunting expeditions. Fishing, on the other hand, is quite evenly spread among both sexes and takes up 33,011 h/annum.

Animal husbandry entails poultry raising only, a minor subsistence activity that is mostly carried out by women or children. Its limited importance is equally expressed in the little time invested, about 15 h/day for the entire community. Chickens forage freely around the house and minor work tasks include the occasional feeding with maize or rice husks or simply shooing chickens away from cooking areas. If cattle were introduced to the community, labour requirements would be distributed differently and the need for male labour inputs would become indispensable. The cultivation of a kitchen garden is primarily a woman's domain; most labour tasks, in fact 76%, are carried out by females and include the planting, watering, weeding and harvesting of fruits and vegetables. Tending a kitchen garden though, is at best a small-scale activity that does not seem to enjoy the same standing as agricultural field work; time inputs are thus correspondingly low.

As far as trading is concerned, agricultural produce is sold to itinerant traders, neighbouring farmers or directly at the market in San Borja. The person-hours invested for trading account for 17,885 h/annum, including travel time. Trading activities have their peak during the annual rice harvest when itinerant traders, in search of harvested crops, visit the community daily. Transactions are slightly higher with men (53%) as they often speak Spanish and generally have more vending experience than women. The involvement of women, on the other hand, should not be underestimated. Apart from carrying the produce to the market, I have mostly seen women engaging in direct door-to-door retailing in San Borja. When honey comes into season, many town-based women eagerly await the arrival of the Tsimane' women in town; freshly bottled wild honey is highly appreciated for its healing qualities and, more importantly, low price. At the time of research in 2004, wage work opportunities were exclusively in male hands, totalling 8,687 h/annum. Expect for the teacher's position, who earns a monthly salary, all wage labour is temporary with an average of 38 working days per household.

The manufacture of tools for the economic system takes up 32,120 h/annum and is done by both males and females of different ages. Under this category, I have counted all preparatory activities for hunting and fishing, as well as handicraft production for the market. As far as the first set of activities goes, we clearly enter into a man's world. The fabrication and manipulation of hunting equipment ranges from sharpening a hunting knife to manufacturing all kinds of bows and arrows. The latter, a particularly favourable chore for rainy days, requires skill and a meticulous eye for detail. Positioning the feathers in order to achieve the best flight path angle is just as important as applying the right quantity of red sticky wax to affix the arrow points to the shafts. Preparing for fishing entails mending the fishing net, for example. As far as handicraft production goes, only the fabrication of those floor

Table 5.3 Labour time investment per person and day in the economic system, Campo Bello, 2004, 2006

		Total labour time/p·cap. (min)	Subsistence labour time/ p·cap. (min)	Cash crop labour time/ p·cap. (min)
Agriculture				
	Rice	32	15	17
	Plantain	21	18	3
	Maize	10	6	4
	Manioc	15	13	2
Housegarden		6	6	0
Manufacture/ handicraft		22	0	0
Hunting		21	21	0
Gathering		20	10	10
Fishing		24	24	0
Animal husbandry		4	4	0
Trading		13	0	13
Wage labour		6	0	0
Total		194	116	49

mats destined for the market was accounted for. The weaving of these mats or woollen bags is exclusively in the hands of women.

Finally, I was interested in the daily time spent in subsistence and market production. To this end, Table 5.3. gives an aggregate account of the total daily labour time invested in the economic system and its further division into subsistence, cash crop and wage labour categories. Notable are the relatively low time investments in cash crop labour, accounting for 25% of total daily labour time and around 40% of subsistence labour inputs. The average daily labour time per person amounts to 194 min, of which the largest fraction is made up of time invested in subsistence production (116 min/person/day or 1.9 h/person/day).[9] Whereas total labour inputs for rice production are highest, they only amount to 13% of all subsistence labour time. Obviously, rice is the most important cash crop, while the export of plantains accounts for much less. As an interesting observation, manioc requires more labour input per person than the production of maize, even though the latter generates more cash when exported. This finding somehow manifests the willingness of the people to invest time in the reproduction of culturally important rites such as manioc beer consumption. On a similar note, the relatively high time investments in fishing and hunting are for subsistence only; again, both activities are also pleasurable pastimes that have a strong socio-cultural meaning attached to them. Finally, time resources invested in wage labour are negligible, accounting for merely 3% of the social system's total labour time.

[9] It may be noted that, despite the fact that the category 'manufacture/handicraft' contains market production, all labour inputs for market production were included in the category 'gathering'.

5.4.2.4 The Community System

We have arrived at the fourth and last functional subsystem: the community system. Only 7% of all person-hours are invested in the reproduction of this subsystem. This is not much when compared to the rest, but somewhat reflects the 'family-level' (see Johnson 2003) society that the Tsimane' may be characterised as. Still, a growing community spirit could be observed in the village and is constantly reproduced by the variety of communal gatherings and events. Public sports and games, for instance, are highly encouraged during school breaks and the community's youths look forward to recurring sports competitions against other communities. Boys and adult men invest more of their time in public games, in fact 75%. Girls, on the other hand, also enjoy the occasional soccer or volleyball game and some are strong competitors for their male counterparts. I found that for females though, this playful manner tends to stop once they turn into an adult; for men, on the other hand, playing ballgames seems socially acceptable for all age groups.

Visiting friends and relatives has become an essential constituent of family and communal reproduction, comprising the largest fraction of time investments in this subsystem. It takes up a total of 85,410 h/annum of both, male and female resources, accounting for 59% and 41%, respectively. The larger time investment of adult males may be related to their more continuous leisure time that provides for longer visits to exchange the latest news. Another explanation may also be rooted in a woman's lack of confidence to make extended visits to neighbouring households which may not form part of the extended family network. One may argue that time invested in visiting kin and neighbours or any other kind of entertainment could also function for the reproduction of the person system. I do, however, consider these activities essential for the reproduction of the community system, as they have a strong socio-cultural meaning within the Tsimane' community network. Time investments in ceremonies and festivals are fairly limited (15,695 h/annum). Village feasts take place regularly but do not take up a lot of time and effort for preparation. Time consumed by religious activities is also rather limited. Most days, people listen to a religious radio programme emitted in their language and during my initial stay in the community, I observed the occasional Sunday mass celebration led by a young Tsimane' who had received training from the New Tribes missionaries.

The final category is 'community and political participation' and accounts for the smallest activity set within the community system (14,235 h/annum). Communal work, however limited in time, has also gained importance in recent years and been taken up by men, while a woman's role is more indirect and limited to the preparation of food. Since the formation of the Tsimane' Council, village politics has gained importance too. The high participation of women in political activities somehow reflects the periods when time allocation studies were undertaken. The first research cycle coincided with intense pre-election campaigning prior to the municipal elections in December 2004 and all present community members were invited to the community house. The second time use study was carried out in the aftermath of a river flood in April and May 2006, when regular village meetings were organised during which household items and food were distributed to the local families. In these distribution processes, women were given an important monitoring role.

5.4.3 *Time, Cash and Efficiencies*

I am now going to bring in cash in/outflows as a means of analysing efficiencies. Upon selling, invested labour time becomes translated into cash which, in turn, gets reinvested in the market. On the inflow side, I will look at the efficiencies of cash incomes from agriculture, gathering and livestock exports, on the one hand, as well as cash incomes from wage work, on the other hand. As far as the latter goes, my main research interest centres on the local strategies to increase the productivity of time.

5.4.3.1 Cash Incomes: Agriculture, Gathering and Livestock Exports

The people of Campo Bello export unprocessed agricultural crops, a small quantity of livestock and a small range of forest products (Tables 5.4 and 5.5). As the main cash crop, rice has two different marketing channels: direct selling in San Borja or via itinerant traders visiting the community. Rice is a high demand product in the region and insufficient rice supplies seasonally drive up its price in the nearby market town of San Borja. Out of a total rice yield of 61.45 t in 2004, rice exports amounted to 32.14 t for the whole village (783 kg per household). Taking the annual average price paid to the farmers (1 kg of rice = US$0.15), the cash generated through the export of rice per family, accounts for an income of 118 US$/annum. The real figures may be even lower, since most Tsimane' households do not take advantage of price fluctuations and export most of their produce shortly after harvest between February and May, when prices are at their lowest (see Vadez et al. 2004, 2005).

In the absence of chemical fertilisers or additional animal labour, the cultivation of rice is still a fairly traditional undertaking in Campo Bello. The only technological 'amplifiers' (see Fischer-Kowalski 2007: 11) are manual rice seeders in order to increase the speed and efficiency of the sowing process. Having said this,

Table 5.4 Cash incomes from agriculture, Campo Bello, 2004

	US$/HH/annum
Rice	118
Plantains	64
Maize	21
Manioc	3
Total	206

Table 5.5 Cash incomes from other sources, Campo Bello, 2004

	US$/HH/annum
Livestock	11
Seed pods	59
Honey, turtle eggs	7
Woven mats	44
Total	110

however, these tools are not used unanimously and some families still apply the traditional cultivation method by using a dibble stick. Depending on the sowing device employed, 1 ha of rice production has been calculated at 108 person-days or 96 person-days of labour input for the entire production cycle (Table 5.6). This means that with the application of rice seeders at the planting stage, a Tsimane' farmer is spared 12 days of hard labour. Plantains, maize and manioc are exported all year round and are much less exposed to price fluctuations. They are frequently bartered for household items and sold more sporadically. Maize and manioc, for example, are repeatedly traded in exchange for beef at a nearby cattle farm.

Cash incomes from gathering have their peak with the marketing of balsa seed pods in September and October. The entire process is a time-consuming endeavour, involving the coverage of long distances in search of the trees, seed picking (at times, high balsa trees are cut in order to reach the seed pods), the transport back home, the extraction of the woolly flesh from the kernels, the weaving of the storage bags and, finally, the arduous journey to San Borja. Woven floor mats account for the second most important source of annual income and were calculated to bring 44 US$ into a Tsimane' household. These mats are produced exclusively by women and frequently sold outside the community. In terms of cash income, livestock, honey and turtle eggs are a minor source only, accounting for less than 5% of total cash earnings.

Table 5.6 Labour days for rice production, Campo Bello, 2004

Activities (days/ha)		Rice production- traditional	Rice production- rice seeder
Field preparation			
	Clearing underbrush	14	14
	Felling trees	21	21
	Burning and cleaning	3	3
Crop management			
	Planting	14	2
	Weeding	28	28
	Harvesting	28	28
Total		108	96

Table 5.7 Wage labour in Campo Bello, 2004

	Total US$	%	US$/HH/ annum	Total days worked	Number of HH
Agricultural farmhand	1,839	40	45	678	16
Cattle ranching	706	15	17	277	6
Road construction	395	9	10	122	3
Logging	175	4	4	54	2
Other	1,516	32	37	379	2
Total	4,631		113	1,510	29

5.4.3.2 Cash Incomes: Wage Labour

In 2004, the village average income from wage labour opportunities amounted to 4.631 US$ for the entire system, accounting for an average of 113 US$ per household (Table 5.7). Most households seek temporary wage labour opportunities as agricultural farmhands in the surrounding area, from where they have the chance to return home regularly and check on their homes and fields. The most intensive agricultural wage labour period, between August and November, coincides with the highest agricultural labour needs in the village. Hence, when peak opportunities overlap, households with a greater number of economically active members are in a better position to diversify and exploit multiple sources of income (see Carlstein 1982). At the time of research, wage labour was exclusively carried out by men. However, a handful of women joined the wage labour market indirectly, as some of the wives accompanied their husbands for longer working periods on cattle farms. We see that men engaging in wage labour opportunities range from 18 to 58, with a peak period in a man's twenties. Work at cattle farms accounts for the second most important wage labour investment, accounting for 15% of all labour inputs. In 2004, a total of six male community members, also older ones, worked for nearby cattle farmers. Often, there exists mutual trust between the Tsimane' family and the cattle rancher, who would visit the community and personally recruit the men. Some men are employed by the same cattle ranchers at regular intervals. Younger men, on the other hand, are usually recruited for temporary labour opportunities in road construction or logging, which both involve longer periods away from the village. Their numbers, however, are much lower; in 2004, only three men were involved in road construction and two adult men found temporary wage opportunities at a logging firm.

In terms of efficiency, we see that most cash is generated through the export of agricultural products. As to the labour input for rice, 1 ha of rice requires about 108 person-days under the traditional method and 96 person-days by using a rice seeder.[10] If an average yield of 1,710 kg/ha is taken, we arrive at an income of 257 US$/ha (1 kg of rice = 0.15 US$), which, in turn, accounts for an average daily wage of 2.52 US$. As an interesting finding, this amount correlates with the daily average amount paid for wage work in the area. If we now look at the working days invested in agriculture and wage work, the question is obvious: Given the low working days in wage labour as compared to agriculture – 37 and 108 (96), respectively – in order to obtain the same monetary return of investment, why do not all Tsimane' households engage in wage labour? The answer seems to be personal preference, Tsimane' informants repeatedly mentioned the exploitative nature of wage work, and older residents particularly proudly affirmed that they never had to work for anyone else but themselves. They also asserted their strong dislike for some outside farmers, who are notorious for their abusive treatment. Younger generations, on the other hand, are more used to interaction with outsiders. Some informants revealed that they were burdened with debts and therefore had to pursue 'quick' wages. They had received credit

[10] As it could not be ascertained how many households actually use a rice seeder instead of a dibble stick (due to frequent borrowing), the average was taken (i.e., 102 person-days) for further calculations.

supplies from travelling traders that had to be paid back within a short period. Wage labour, nonetheless, is not a phenomenon young men tend to engage in uniformly, but rather an individual choice. If a man is well embedded in a tight family network comprising many economically active members, he is somewhat free to choose what he knows best, without the innate need for additional wage labour.

5.4.3.3 Cash Investments

If the way to the intensification of production lies in a gradual increase in the average length of a working day, a society is faced with two options: expanding the time resources available in the system, or intensifying the use of time through efficiency gains. In light of Campo Bello's demographic growth rate (see Boserup 1965), the community is faced with a similar choice. Thus examining the community's current strategies to increase the productivity of time is what this section entails. One way of doing so is through the lens of cash investments. What we see is that substantial amounts of cash are indeed spent on technological 'amplifiers' like rice seeders and agricultural tools to relieve human labour inputs in agriculture (Table 5.8). All households spend money on basic tools that are necessary for agricultural production. Without the indispensable machete, for example, no fields could be cleared or plantains harvested. The introduction of rice seeders has cut the work-load tremendously during the planting stage, in fact by almost 90%. On a similar note, while the traditional rice cultivation method requires the input of more people, manual rice seeders are employed by one person only. At the time of research only 23 households were in possession of such a device. Yet contrary to other tools, rice seeders are frequently passed between families, even if not related by blood. To increase the productivity of hunting labour time, shotguns and rifles have largely taken over from the use of more traditional hunting devices. Nylon fishing nets were encountered in 21 households at the time of field study. With such a net, substantially more fish can be caught in a much shorter time frame. If we look at transport, the use of a bicycle has become a popular means of commuting within and between settlements. Rather than solely being a means of transport for oneself, it equally provides the opportunity for transporting agricultural products (e.g., rice and plantains) to the market, thereby also increasing the efficiency of travel time.

If we look at the owners of these commodities, we find all of them in the hands of men. In 2004, all locally owned bicycles were in the possession of male residents, some of whom would spend considerable time and effort in repair and maintenance. Similarly, rice seeders, fishing nets or a shotgun are also exclusively purchased by male residents. The same goes for the purchase of status symbols like a watch, a radio or a pair of shoes. It was generally observed that women tend to purchase commodities that are of use for the reproduction of the household system, such as wool for bag weaving, hygiene articles, general household appliances, or simply food.

I finally found a correlation between the source of earnings and spendings. Itinerant traders, for example, frequently offer barter products in exchange for agricultural products. As these traders often receive their trade goods on credit from

Table 5.8 Cash investments in Campo Bello, 2004

	Number of items	US$/per item	Total US$	Function	
				Time-saving	Socialisation
Labour amplifiers					
Rice seeder	23	5.4	125	X	
Machete, axe, spade	176	6.0	1,063	X	
Shotgun	34	18.1	616	X	
Fishing net	21	10.9	228	X	
Transport amplifiers					
Bicycle	17	21.8	370	X	
Status symbols					
Radio	34	9.7	329		X
Watch	24	7.3	174		X
Clothes, shoes	n.a.	6.0	302		X

shopkeepers in San Borja, these kinds of transactions hold the greatest probability of exploitation for the Tsimane'. One *arroba* of rice (11.4 kg), for instance, is sometimes traded for one litre of alcohol or a handful of radio batteries. Other barter products include pasta noodles, sugar, salt, fish hooks, medicines and aluminium pots for cooking. As a general observation, barter products for agricultural exports tend to constitute commodities for every day use. Cash earnings from wage labour, on the other hand, are likely to be spent on status goods like a radio, a watch or a bicycle. This relates to the type of payments received, as they are frequently paid a lump sum at the end of their labour commitment which can be readily invested in more expensive items. At the same time, it is not common practice for a Tsimane' man to send remittances home while away on wage labour activities.

5.5 Conclusion: Discussing Opportunities and Time Squeeze

This chapter has analysed Campo Bello's human time use along the following lines: A calculation of the system's socially disposable labour time, an analysis of the person-hours invested in each functional subsystem, and an examination of cash flows and efficiencies. What have we learnt from this analysis in terms of the system's opportunities, time squeeze, and other bottlenecks?

Concerning the first analysis, the advantages of examining disposable labour time by age/sex group at the system level, rather than focussing only on the individual or household level, have become evident at various stages throughout the discussion. The potential contribution of labour time may be highest in an age group that is free from physical, social and educational constraints. What we see for Campo Bello, however, is the limited availability of people in their most productive years. On the other hand, the system comprises extremely high numbers of children, a rather typical feature of agricultural societies. What does this mean in terms of disposable labour time and time demands on others? Cain's (1980: 212) description may be an instructive point of departure. In a child's life-cycle, he distinguishes four different phases from a child's complete dependence on parental time, to becoming economically more active but producing less than he or she consumes, followed by a period during which the child produces more than he or she consumes to finally becoming as equally productive as an adult. Productivity increases, so he further contends, depend on age, working time, the efficiency and the productiveness of the tasks performed. In Campo Bello, we see that the age group with the highest numbers is the 0–5 year old range, who do not contribute to labour time but are economic dependents of their parent's, especially their mothers', time. Nonetheless, there is evidence that the more children are added to the family, the less are the time costs of child rearing on the mother's time resources, since older children can step in and take over some of the chores. By the age of 5 or 6, children of both sexes become economically active, engaging first in skillfully undemanding 'enabling' labour, obviously producing less than they consume. These labour tasks are mostly invested within the household sphere and, interestingly, the culturally prescribed division of labour does not appear too stringent during this period. As boys gradually move away from their mother's sphere of influence around the age of 8 or 9, they start to become more actively involved in directly productive labour. They may occasionally bring home some fish or help their fathers in the field. Girls, on the other hand, seem to start dividing their disposable labour time between household chores and more productive tasks like gathering or fishing. Both sexes, however, lack physical maturity and are constrained by educational demands that have become strongly embedded in the local culture in recent years. For many girls, the transition from being a pupil to becoming a wife and a mother seems fairly abrupt. For men, they often spend their most productive years after school temporarily away from the community as wage labourers. With the absence of these community residents, others have to increase their work-load, adding to their own labour demands. This is particularly felt by young, married women, who, in the absence of their husbands, have to become even more active in the productive sphere in order to ensure their family's daily reproduction needs. For some activities, this would mean increasing her work day; often to the detriment of sleep or other functions for personal reproduction. However, due to physical and social constraints, not all activities normally provided by the husband, can be done by a woman (e.g., hunting). For these periods, the existence of a tightly woven extended family network to fall back on, becomes crucial.

Let us now turn to the individual subsystems. What are the findings in terms of time squeeze and opportunities for change? To start with the person system, it does not seem to offer much leeway for changes in time use. Functions for basic reproduction like sleeping and eating are both activities that cannot be delegated to others. It was an interesting discovery that bodily regeneration seems to be more socially acceptable for adult men than for women; a finding that pertains also to the elderly generations in the village. When it comes to hygiene, men can invest their time resources for their own hygiene, whereas women are further responsible for the hygiene of their infants. Still, I found that men use up more of their daily time for personal hygiene than women. This may be related to the fact that men tend to enjoy longer bathing sessions after a day of hard work, an activity that also has a strong leisure component to it. Overall though, time resources are divided fairly equally between the sexes and the time allocated for basic reproduction appears sufficient when compared to other, more modern societies. Wouldn't the disposal of a 3-h period of rest and complete inactivity every day be an appealing thought for us living in the industrial world? The remaining 10% of all time inputs into the person system is taken up by extended personal reproduction, comprising studying and education. This takes up about 4 h every day, a time that could otherwise be used for fishing, light agricultural work or simply playing. But parents are well aware of the benefits of schooling and, with the knowledge of Spanish and arithmetic skills, children will be less likely to be exploited by future employers. The introduction of school classes has also changed a child's time demands on the mother. For a 4-h period every day, mothers are free to engage in a range of activities that would not have been feasible otherwise. At the same time though, children cannot contribute to other tasks during these morning hours. Time invested in schooling is hardly likely to increase, since further educational opportunities in the village are limited. Sending one's children away for higher education is not common practice. In fact in 2004, only the teacher's son attended high school in San Borja, where he could stay with a relative.

Concerning the reproduction of the household system, female time investment is a little over 60%. This contribution would be higher in societies, where the sexual division of labour manifests itself stringently after socialisation begins (see, for example, Cain 1980). In Campo Bello, however, little boys also help with domestic chores and care for dependents, thereby diminishing the overall female contribution. A closer look reveals, however, that women of all ages bear the main burden for both, the day-to-day running, as well as, though to a lesser extent, the long-term maintenance of the domestic unit. As an interesting footnote, time demands on women have, nonetheless, become less over the past decades. Up until the 1980s, women would have to provide for their families' clothing, a time-consuming manufacturing process, during which the bark of a certain tree was dried and processed into white cloth. Today, all clothing is traded on the market. The same goes for cotton, which, not long ago, would be processed from natural fibre and used for the weaving of cotton bags. This lengthy process came to an end with the advent of commercial wool on the regional market.

If we now look at overall time investments, are time demands on the household likely to change in the future? To start with, care for dependents takes up most of a woman's time resources and, in light of demographic growth rates, is likely to increase. On my last visit to Campo Bello in early 2008, the local population had grown to 250 people (as compared to 235 in 2006 and 231 in 2005), for the most part as a result of high birth rates. This growth rate was also reflected by the growing number of children attending school; 50 pupils were registered, as compared to 38 and 35 in the years before. It seems that having many children provides an asset at two levels; first in terms of additional labour, second in terms of social standing within the community. Whereas no formal institutions exist to 'outsource' the work of child care, as there would be in modern societies, the presence of elderly women in the household has an enormous impact on the work-load of a young mother. As my empirical data reveals, older women contribute 2.2 h/day to the care of dependents. Again, we see why a closely-knit family network is of so much importance to a Tsimane' family. In light of steady population growth, time investments for food production are also likely to increase. In the absence of gas cookers or rice peeling machines, both of which are unlikely to arrive in the village in the near future, food preparation will continue to take up most of a woman's morning hours. As far as the production of beer is concerned, we may observe a slight decrease in production, as the community is becoming increasingly integrated into the regional socio-economic culture. At official village feasts, for example, the consumption of purchased alcohol has become much more popular. Within the confines of the household, however, this change has not been observed, at least so abruptly, and beer preparation still takes up many of the female time resources available in the household. As far as house building goes, all individual dwellings are still exclusively built from traditional materials that have a life span of 4–5 years. Obviously, a change to more resistant building materials would spare a man's time in maintenance and repair work. Yet more resistant materials would mean the investment of additional economic resources, which are currently not available to most families. Finally, given the lack of a road to the regional market, the transport of bulky material would add a further constraint.

In light of the household being a predominantly female space, wouldn't we expect the economic system of Campo Bello to be mostly in male hands? In our case though, males contribute only slightly more of their time resources, namely 53%, while female residents account for the remainder. Women are active economic producers, even if their activities are mainly confined to agricultural and horticultural activities, fishing and gathering. Agriculture requires the most time resources and people of all ages have their share of the agricultural work-load. Men insert some 2.6 h/day in agricultural tasks and are in charge of most energy-intensive labour tasks such as slashing and burning. People still use fairly traditional agricultural methods and only human labour is inserted into reproducing the agricultural system. The introduction of technological 'amplifiers' such as rice seeders to accelerate the sowing of rice, has been a rather recent phenomenon within the community. It was an interesting observation that these time-saving devices are freely circulated among the community members, while this is not the case with other market goods. Although not empirically grounded, this may be an expression of a strong communal sense of solidarity to ensure that there exists the same ceiling on the average expenditure of labour for everyone (see Descola 1996: 294). Using this technology saves not only days of work, but further reduces labour requirements within a household. Thus a task that previously took the efforts of two to three people can now be done by one man in a much shorter period. A person invests only 116 min/day (1.9 h/day) in subsistence labour and this is rather low compared to other agrarian regimes. We also see that for some cultigens, the return upon investment is fairly low. This, however, is compensated by the socio-cultural pleasures that people derive from these labour inputs. The production of maize requires one third less labour that the production of rice and about half as much as the cultivation of plantains. Moreover, local farmers usually plant maize twice a year and harvesting can take place at more relaxed periods of the agricultural year. Adding to this, maize as a cash crop does not undergo such drastic price fluctuations as rice, thereby assuring steady cash returns all year round. Despite these advantages of maize over rice, we may note that the production of rice still exceeds the production of maize by a factor of five. From what I have gathered though, rice enjoys a much higher standing over maize, something that is also reflected in the wider regional economic system. On the regional market, rice is in demand all year round and a bag of rice is easily traded in exchange for pasta noodles.

Overall, we see that Campo Bello only invests 13% of all available time resources in the reproduction of the economic system. Would there be room for expansion or intensification of labour inputs in certain economic activities? To this end, I would like to concentrate on agriculture, on the one hand, and wage labour, on the other. At present, the exertion of agricultural tasks relies solely on human labour and the introduction of additional animal labour would not be a feasible option. Animal labour is neither culturally ingrained in their practices, nor is it practiced at higher scale levels (e.g., by non-indigenous farmers in San Borja). The expansion of labour time would hence be one of the options for increasing agricultural outputs. Let us first scrutinise the economic system in search of possibilities to shift labour time from other productive activities. What we see is that labour investment in agriculture takes up less than half of all labour time inputs within this subsystem. Fishing takes up one third of all time inputs, followed by hunting and gathering. Simply shifting time resources would not be feasible since most agricultural tasks require physical capabilities and capacities that cannot be provided by children or women. Fishing, for instance, may require some skill, yet does not involve the use of heavy equipment. Similar to gathering, these activities are also suitable for children. A boy invests some 0.8 h/day in agriculture, while a girl's labour time input accounts for 0.2 h/day. If we consider that most of a child's day is taken up by studying, playing, child care, rest, eating, fishing and gathering, then time resources for agricultural labour for this age stratum remain fairly limited. Likewise, some agricultural tasks call for a certain skill level as well as physical stamina, both of which boys and girls may lack. Reaching beyond the economic system, one of the possibilities (except for reducing some of the leisure time in the person system) would be to decrease time investments in the community system. This would mostly impinge on visiting and socialising with kin and neighbours, an activity that is deeply ingrained in the local social fabric. Also, the expansion of the agricultural work day may mean a reduction of sleep at agricultural peak times that would be felt by adults particularly. Nonetheless, a likely expansion of agricultural production time is hampered by limited marketing facilities. For the time being, Campo Bello remains largely remote and lacks the adequate transport facilities to market centres. At various times of the year, the people have to rely on the market coming to them rather than vice versa. As far as intensification is concerned, the main constraint is the lack of economic resources. Chainsaws, for example, would provide a relief on labour demands when opening a rice field, whereas the use of chemical fertiliser would ease labour inputs in weeding. At both periods of research, however, no residents were in a possession of a chainsaw and only one younger farmer used a small amount of chemical fertiliser.

Wage labour is the second economic activity that I would like to examine closer in respect of the scope for time use expansion. At present, labour inputs into wage labour account for merely 3% of total time inputs into the economic system and, at least at the time of research, was exclusively a male endeavour. Wage labour opportunities are highest during the rice production period that peaks between September and November. This period, however, coincides with one's own field preparation and rice cultivation. Hence, while unmarried adolescents are somewhat freer to use their time resources working on someone else's field, this does not apply to those with a family. In the absence of one's own newly opened rice field, a family has to fall back on the family network for the consumption of rice. More importantly though, lacking rice means lacking cash currency for purchasing pasta noodles and other household artefacts necessary for personal and household reproduction. Wage labour opportunities for women in the area are limited and with regard to our community it seems improbable that this phenomenon, apart from occasional opportunities, will occur. First, women are indispensable for the daily food provision and the general reproduction of the household. Therefore, longer periods away from home would only be possible for those households with more adults. Another aspect to be taken into consideration is the fact that the lack of Spanish language skills may impede women entering the labour market. In terms of time and cash flow efficiencies, the most interesting finding is that the marketing of rice generates similar cash returns similar to wage labour. Rice production requires an average of 102 working days, while average time investment in wage labour amounted to 38 days only. Adding to this time advantage, wage labour opportunities are more flexible and residents have the liberty to allocate their time resources freely, to a certain extent. Most wage labour, however, is notorious for being exploitative. Older informants expressed a strong dislike for wage labour and in 2004, two-thirds of all local households had in fact never engaged in this kind of labour.

Time invested on the functioning of the community system accounts for only 7% within the whole system. Still a 'family-level society', there are signs that communal interaction is becoming increasingly more time-consuming. In parallel to the more traditional activities, recent cultural adaptation to national society has resulted in the celebration of national holidays, birthdays and regular village feasts. In addition, sport events have become more frequent, and are mainly a result of the teacher's initiative. As a general observation, therefore, an ever growing sense of community and the increasing exposure to the regional market economy are likely indicators that time demands for the reproduction of the community system will increase.

Finally, what do the local people spend their money on? It appears that people only invest part of their cash in time-saving assets. Besides investing in rice seeders, every household possesses a variety of agricultural tools without which certain tasks would simply not be possible. Most men own shotguns or rifles that have by and large made the elaborate and time-consuming manufacture of bows and arrows obsolete. Likewise, the use of fishing nets enables people to intensify their time resources, as substantially more fish can be caught in a much shorter period. Adding to this are intensification strategies to increase transport means and levels. First, in 2004 around 10% of all male residents in Campo Bello were in the possession of a bicycle. To my mind, the reasons why women do not own bicycles can only indirectly be traced back indirectly to socio-cultural barriers. Most evidently, women's cash incomes from selling garden and forest crops are lower than the lump sum a man would earn from wage labour. But even if a woman receives a lump sum of cash, it would probably not be socially acceptable to spend it primarily on herself rather than sharing it with her family. Interestingly, one quarter of all functional items purchased on the market have a socialisation function, such as watches or radios. What can be observed is the phenomenon of 'time aggregation', rather than investing in inputs to release the people from confining time demands. This may be partly explained by the fact that people may not feel immediate time constraints. They invest rather low labour inputs to meet subsistence needs and rely on a variety of production modes. As shifting cultivators, they only have to invest in a small number of indirect and labour-intensive activities to render the natural environment useful for their needs. Likewise, with the back-up of an extended family network, individuals may engage in those subsistence activities they like most. Good hunters are respected for their skills and there is always the possibility to exchange game meat for garden crops.

References

Bernard, H. R. (2006). *Research methods in anthropology: Qualitative and quantitative approaches* (4th ed.). USA: Altamira Press.

Bevans, G. E. (1913). *How working men spend their spare time*. New York: Columbia University Press.

Boserup, E. (1965). *The conditions of agricultural growth. The economics of agrarian change under population pressure*. Chicago: Aldine/Earthscan.

Cain, M. T. (1980). The economic activities of children in a village in Bangladesh. In H. P. Binswanger (Ed.), *Rural household studies in Asia* (pp. 218–247). Singapore: Singapore University Press.

Carlstein, T. (1982). *Time resources, society and ecology, Volume 1: Preindustrial societies*. London: George Allen & Unwin.

Descola, P. (1996). *In the society of nature: A native ecology in Amazonia*. Cambridge: Cambridge University Press.

Ellen, R. (1982). *Environment, subsistence and system: The ecology of small-scale social formations*. Cambridge: Cambridge University Press.

Firth, R. (1967). *Themes in economic anthropology*: A general comment. In R. Firth (Ed.), Themes in economic anthropology (pp. 1–28). London: Tavistock Publications Limited.

Fischer-Kowalski, M. (2003). *Socio-ecological regimes, time use and the environment*. Presentation at the second international conference of the International Society for Industrial Ecology (ISIE), hosted by the University of Michigan, Ann Arbor, 29th June – 2nd July 2003.

Fischer-Kowalski, M. (2007). *Ageing, time use and the environment*. Presentation at the workshop: Research foresight for environment and sustainability – megatrends and surprises, EEA Copenhagen.

Fischer-Kowalski, M., Singh, S. J., Ringhofer, L., Grünbühel, C. M., Lauk, C., & Remesch, A. (forthcoming). Sociometabolic regimes in indigenous communities, in scale interactions with an industrial world context, and the crucial role of working time. A comparison of case studies. Submitted to *Human Ecology*.

Gershuny, J. (2000). *Changing times: Work and leisure in post-industrial societies*. Oxford: Oxford University Press.

Goffman, E. (1959). *The presentation of self in everyday life*. London: Penguin.

Harvey, A. S., & Pentland, W. E. (1999). Time use research. In W. E. Pentland, A. S. Harvey, P. M. Lawton & M. A. McColl (Eds.), *Time use research in the social sciences* (pp. 3–14). New York, Boston, Dordrecht, London, Moscow: Kluwer Academic/Plenum Publishers.

Johnson, A. (1975). Time allocation in a Machiguenga community. *Ethnology, 14*, 310–321.

Johnson, A. (2003). *Families of the forest: The Matsigenka Indians of the Peruvian Amazon*. Berkley: University of California Press.

Netting, R. M. (1993). *Smallholders, householders. farm families and the ecology of intensive, sustainable agriculture*. Stanford: Stanford University Press.

Pember-Reeves, M. (1913). *Round about a pound a week*. London: Bell.

Piland, R. A. (1991). *Traditional Chimane agriculture and its relation to soils of the Beni biosphere reserve*. Bolivia: Master thesis, University of Florida.

Pimentel, D., & Pimentel, M. (1979). *Food, energy and society*. London: Edward Arnold.

Rappaport, R. A. (1968). *Pigs for the ancestors: Rituals in the ecology of a New Guinea people*. New Haven: Yale University Press.

Sahlins, M. (1972). *Stone age economics*. London: Tavistock.

Salisbury, R. F. (1962). *From stone to steel: Economic consequences of technological change in New Guinea*. London: Cambridge University Press.

Sieferle, R. P. (2001). *The subterranean forest. Energy systems and the Industrial Revolution*. Cambridge: The White Horse Press.

Sorokin, P., & Berger, C. (1939). *Time-budgets of human behaviour*. London: Harvard University Press.

Szalai, A. (1972). *The use of time: Daily activities of urban and suburban populations in twelve countries*. The Hague: Mouton.

Vadez, V., Reyes-Garcia, V., Godoy, R., Apaza, L. V., Byron, E., Huanca, T., et al. (2004). Does integration to the market threaten agricultural diversity? Panel and cross-sectional data from a horticultural-foraging society in the Bolivian Amazon. *Human Ecology, 32*(5), 635–646.

Vadez, V., Reyes-García, V., Huanca, T., & Leonard, W. R. (2005). *Cash cropping, farm technologies and deforestation: What are the connections?* Tsimane' Amazonian panel study working paper 18, 18th of December 2005.

White, B. (1976). Population, involution and employment in rural Java. *Development and Change, 7*, 267–290.

Chapter 6
Comparing Local Transitions Across The Developing World

Abstract In this chapter the local system of Campo Bello will be compared to three other local systems from different world contexts. What they have in common is the fact that they are quite 'remote' in terms of accessibility and 'backward' compared to the general economic development of their country. The aim of such a comparative undertaking is to find out whether the courses of development taken by each of these local settings actually follow a similar pathway; and if this is the case, what can be learnt for a sustainability transition? Specifically, I will look at three different sets of biophysical indicators – ecological, agro-economic and human labour time – in order to find whether the systems, which are all at different sociometabolic transition stages, undergo, in biophysical terms, similar transformation processes. The second specific aim is to examine some of the drivers for change. The whole discussion is embedded in two theoretical frameworks: the intensification theory and the sociometabolic regime theory.

6.1 Introduction

A society's interaction with its environment may be driven by the specific dynamics unique to the individual setting. At the same time though, we may well observe similar processes in different contexts, irrespective of time or geographical setting. Analysing these common patterns from a biophysical perspective is what this chapter is concerned with. In so doing, I hope to do justice to the second aim of the book, namely to enhance our understanding of local sociometabolic transitions – a process that is ongoing in large parts of the world today. To this end, this chapter comprises a comparative approach; it provides a 'snapshot' of four contemporary local social systems that are all, though at different stages and with different intensities, undergoing a transition. Whereas all cases are situated in different developing world contexts, what unites them is the fact that they are quite 'remote' in terms of accessibility and 'backward' compared to the general economic development of

L. Ringhofer, *Fishing, Foraging and Farming in the Bolivian Amazon: On a Local Society in Transition*, DOI 10.1007/978-90-481-3487-8_6,

their country. However isolated and self-contained they may be, contemporary local systems at the beginning of the twenty-first century interact, though with different intensities, with the outside world and are equally impacted upon by developments occurring at higher scale levels (i.e., at the regional, national or even the international level). Hence, as they no longer face the resource constraints of pre-industrial sociometabolic regimes, the specific aim of such a comparative undertaking is to describe these contemporary courses of development. To what extent are their transition pathways driven by their own process dynamics, or can we also observe similar transition patterns? And with these insights, what can be learnt for a sustainability transition?

To say that this chapter was inspired by Clemens Grünbühel, Simron Jit Singh and Marina Fischer-Kowalski would be an understatement, as it is every bit the idea of their recent scientific efforts somewhat adapted to my own needs. Following their publication (Grünbühel et al. 2007), ideas were further developed during various follow-up meetings, from which a further publication is about to emerge (Fischer-Kowalski et al. forthcoming). The biophysical indicator set has been expanded to now include a detailed time use study for Nalang, Campo Bello and Trinket.[1] Christian Lauk's influence on theoretically combining intensification theory with the theory of sociometabolic regimes is visible in Fig. 6.1. Hobbes' (2005) research and reflections served as an additional source of inspiration with respect to the food security and market dependency indicators in Section 6.3. The notions presented in this chapter, although I alone take responsibility for any deformations that may have occurred, can therefore quite appropriately be called a product of this fusion of ideas and inputs.

Based on the same goals, theoretical frame and a similar general set up as Grünbühel et al.'s (2007) publication, the main interest of this chapter centres on analysing how local systems organise their biophysical flows with their environment at times when industrial transformations taking place at higher system levels produce changing framework conditions. The two explicit goals are to first of all compare the biophysical indicators of the four cases in anticipation of detecting any patterns according to how far the systems have moved forward within a transition stage. The second specific aim is to examine some of the drivers for change. To this end, the discussion opens with a qualitative introduction of the three new local sites providing a brief historical, ecological and socio-economic sketch. Two theoretical frameworks provide the basis upon which to enable a subsequent comparison of biophysical transition indicators: intensification theory and the theory of sociometabolic regimes. Both are combined in Fig. 6.1. which offers a matrix allowing for a sequential order of the local systems according to their modes of subsistence and market integration. Next follows the assessment of three sets of quantitative biophysical parameters – ecological, agro-economic and human labour time – in order to improve our understanding of the biophysical changes taking place during differ-

[1] For SangSaeng, there is no comparative time use data available, since the generation of time use data is largely restricted to labour investments in agricultural activities and hence does not suffice to establish a 24-h time use profile for the whole system.

ent transition stages. When the transition indicator analysis draws to a close, I will move to discuss some of the drivers causing a sociometabolic regime to change. What new options open up, and which environmental pressures get under way?

The four examples with which I try to illustrate different transition stages are no doubt an insufficient basis upon which to construct a general rule. However, it is still hoped to at least observe a trend that, of course, would need further reassessment in a future research endeavour.

6.2 Introducing SangSaeng, Nalang and Trinket[2]

The three sites, SangSaeng in Thailand, Ban Nalang[3] in Laos, and Trinket, an island in the Nicobar archipelago belonging to India, are all rural settings located within the region of south and southeast Asia. All data presented on SangSaeng and Nalang was generated by Clemens Grünbühel. Concerning SangSaeng, the data was collected in the course of two 6-month field studies in 1996 and 1998. As to Nalang, all data was generated during various field studies between April 2001 and February 2002. With respect to the case study of Trinket, data collection was done by Simron Jit Singh during various sojourns on the island between 1999 and 2007. Field work largely took place during the dry season between October and April.[4]

6.2.1 Site One: SangSaeng

The village of SangSaeng is situated in northeast Thailand, in the south-eastern part of the province of Isan near the borders of Laos and Cambodia. With a population of 171 (in 1998) and a village size of 184 ha, its population density is 93 cap/km^2. The village is characterised by sandy soils and was formerly covered by semi-deciduous rainforest. Today, very little remains. Annual precipitation amounts to approximately 1,400 mm. Similar to the entire Isan plateau, the area is flat and there are no continuous natural waterways nearby, with the only exception of the Mun river that passes some 20 km away from the village.

As a part of the Thai-Lao cultural and linguistic area, the Isan are ethnic Lao and display very similar traits to the majority population of the neighbouring People's

---

[2] For a detailed description on SangSaeng see Grünbühel et al. (1999, 2003), on Nalang see Mayrhofer-Grünbühel (2004) and on Trinket see Singh (2003a), Singh et al. (2001) and Singh 2003b).

[3] Hereafter, Ban Nalang will solely be referred to as Nalang.

[4] To assess the biophysical parameters of these three communities the same methodology was followed as described in Chapter 2 of this monograph. For a more detailed discussion on data collection methods see Singh et al. 2001; Singh and Grünbühel 2003; Grünbühel et al. 2003; Singh and Schandl 2003; Schandl and Grünbühel 2005.

Democratic Republic of Laos. Despite national integration efforts, the twenty-one million inhabitants of Isan province still live according to strongly rooted traditions and beliefs. They still speak Lao and maintain a lifestyle that is very much connected to their natural environment. Like most other rural Thais, the Isan are Buddhists and engage in subsistence rice farming. As Isan society is little stratified, SangSaeng can also be described as socially homogenous (see Grünbühel et al. 1999). Having said so, people do, however, differ in the respect they evoke. Successful labour migrants, for example, generally enjoy a higher social status among community members due to their ability to trigger the influx of market goods. Except for these differences, relationships of reciprocity and a culturally embedded sense of solidarity are still prevalent features of daily communal life. The local religious belief system merges Buddhist faith with a spiritual worldview. Spiritual beings are both, harmful and beneficial, and knowledge of how to appease them is an essential component for every interaction of the people with their natural environment. While land is privately owned, forest game and plants (except for fresh wood) are free resources for all villagers.

Due to its geographic remoteness, the Isan province does not have any major economic centres and, except for the migrant workers, is little integrated into the rising national economy of Thailand. Adding to this, the region is characterised by relatively low soil fertility and a high level of deforestation. The people of SangSaeng are engaged in an economy that revolves around three main axes: agriculture based on rice-farming, foraging and labour migration. Rain-fed paddy rice cultivation comprises the central pillar of the people's subsistence economy. Glutinous rice is the most important constituent of the Isan diet and normally consumed at least three times a day, with only a small fraction of harvest surplus sold on the market. Cash crop rice is usually of the state-promoted varieties, which the producer families hardly consume themselves.[5] For the production of rice, people apply fertiliser and make use of electrical water pumps, ploughs and motor-ploughs. Besides rice, other commercial production in rather limited. Families own vegetable gardens and keep chicken and ducks exclusively to provide food for their own consumption. Buffaloes are raised as working animals, meaning that only cattle are destined for the local market. Hunting and gathering constitutes the pillar of SangSaeng's village economy. These activities help to diversify the diet, especially during the dry season, when gardening is limited or made virtually impossible. A wide variety of animals, such as fish, birds and insects, as well as different kinds of leaves, herbs, mushrooms and roots are extracted from the fields and adjacent forests.

Off-farm wage labour constitutes SangSaeng's third main economic pillar. Temporary migration to urban centres or coffee and rubber plantations in southern Thailand has become a common strategy for most locals, especially the younger generations. In most cases, labour migration is inherently linked to the village economy: migrants supply their families with regular monetary remittances and take rice with

[5] Only 5% of Indian rice produced in SangSaeng is consumed within the village. In comparison, 90% of glutinous rice production is consumed by the producers themselves within a year (Grünbühel et al. 2003: 60).

them to their location of work. However, while migrant workers tend to spend the low-work dry season on wage labour outside the community, they tend to return to SangSaeng for the peak work season of rice production. These peak periods, comprising the preparation and transplantation of rice nurseries, require all able-bodied family members to participate either in the direct production process or, more indirectly (especially the elderly and children), through the provision of enabling labour.

6.2.2 Site Two: Nalang

The Laotian community of Nalang lies in the district of Fuang, Vientiane Province, at a distance of about 200 km from the capital city Vientiane. Comprising an area of 1,630 ha, Nalang has a total population of 702 (2001); the population density thus being 43 cap/km^2. Due to its fairly mountainous topography, large stretches of the available land cannot be converted for permanent agricultural use. The main land cover types include 'evergreen forest' in the lowlands, 'high density mixed deciduous forest' in more inclined areas, and 'limestone rock' for the Nalang landmark, the *Phalang*, located in the centre of the village. Annual precipitation rates for the entire area amount to approximately 1,500–2,000 mm (Mayrhofer-Grünbühel 2004: 142).

Historically and especially in the aftermath of the Indochina War in 1988–1989, various immigration waves have affected the ethnic composition of the village. After an influx of Phuan migrants shortly after the war, the village has witnessed rather gradual immigration waves of Kh'mu who, in the past, would engage in shifting cultivation. Migratory flows have resulted in social differences in the village make-up as original settlers tend to own more land and more material resources than more recent migrants. Interestingly though, social discrimination can also be observed among the immigrants. Whereas the Phuan, due to similar ethnic traits, have integrated more easily in the existing village structures, Kh'mu families have remained on the social margins of the community. Mayrhofer-Grünbühel (2004: 94) argues that it was due to the absence of suitable agricultural land resources that more recent immigrants had to revert to shifting agriculture on relatively infertile land. As an interesting footnote, all landless households and those without paddy fields in Nalang are ethnic Kh'mu.

Although Nalang is a Buddhist community, spiritual beliefs are strongly ingrained in the local socio-cultural environment. Common rituals in June and November are generally performed to pacify the village spirit *phi ban*. Both fields and individual homes are dwelling places for spirits and people have the obligation to keep them entertained.

Nalang portrays a subsistence economy dominated by rice that is cultivated mainly in permanent paddies and, to a lesser extent, in swidden fields. This is a typical feature of the whole region, where rice farming represents the main agricultural activity for rural producers. Like most Laotians, the local people produce primarily glutinous rice (*Oryza glutinosa*) that is cultivated in the wet season only and consumed locally. In very productive years, a harvest surplus may either be sold to other villagers or to outside traders. Paddy fields are invariably located at the bottom of valleys and are all fed by a network of free-flowing irrigation channels. Besides rice cultivation, gathering, fishing and hunting activities are also common pursuits the villagers engage in. Foraged foods like bamboo shoots, a great variety of plants, roots, tubers, palms, insects and mushrooms are a welcomed variation to the rice-based local diet. The ingestion of game and fish, on the other hand, provides the people with the necessary protein intakes. To catch fish, people apply a variety of techniques ranging from line fishing to casting nets. Hunting is carried out either by using traditional hunting devices like traps or bows or more elaborate home-made guns. In former times, the rearing of buffaloes comprised another significant feature in the community's agricultural system. They were used mainly for the ploughing of fields and general transport, but also had a strong cultural meaning and worth, as their meat was regarded as a ceremonial food during rituals and village festivities. The possession of buffaloes also meant an additional asset for families, and was widely used as a currency for dowries. Since the arrival of the motor-plough in the mid-1990s, however, the agricultural need for buffaloes has become somewhat diminished. For meat production, buffaloes are gradually being replaced by cattle, largely because maturing times are more rapid. In an attempt to facilitate timber extraction in the area, a road was constructed in 1980, connecting two important market centres in the area. In order to reap the benefits of such a mode of transport, the people of Nalang relocated from the original site on the banks of the Namlang River closer to the road.

With now greater accessibility and increasing market needs and wants, the production of cucumber was introduced as an important dry-season cash crop in the late 1990s. The cultivation of cucumbers takes place after the annual rice harvest, when small cucumber gardens are created along the waterways on the area of the recently harvested fields. Cucumbers are grown for sale on local markets and require relatively large manual labour inputs. Apart from cucumbers, the less labour-intensive cultivation of bananas provides an alternative source of income for the people of Nalang. Greater ease of transport has also had an impact on the extraction of natural resources from the forest. Timber is logged for village use as well as for sale to traders. Increased market integration has further instigated an influx of agricultural machinery and, in so doing, brought about some changes in the local agricultural system. Despite growing market integration, however, subsistence production still remains the most important economic activity in Nalang. People depend little on traded food items and dwellings are built without much dependence on imported raw materials. Still today, the villagers have a profound knowledge of their local environment and use of natural resources.

6.2.3 Site Three: Trinket

Trinket is different to the previous two cases as it is one of the 24 islands that constitute the Nicobar archipelago in the Bay of Bengal. The Nicobar Islands are home to an outstanding tropical biodiversity. Due to annual monsoons, precipitation rates are fairly elevated ranging from 3,000 mm to 3,800 mm each year. The vegetation of the islands is divided into the coastal mangrove forests and the interior evergreen and deciduous tropical and subtropical moist broadleaf forests and extensive interior grasslands. Most islands are surrounded by coral reefs that prevent easy accessibility. Trinket Island is located some 1,200 km off the east coast of India, with an area of 3,626 ha. Sine 1956, the Nicobar Islands fall under a special regulation, the Andaman and Nicobar Protection of Aboriginal Tribes Regulation, which strictly regulates entry. Thus, contacts with outsiders have been restricted to government officials and mainland traders. The inhabitants of Trinket, just as their counterparts on the other islands, live in communities along the coast protected by natural bays and mangroves. At the time of research in 2000, the total population of Trinket consisted of 399 inhabitants living in 43 households spread over three villages. This gives a population density of 11 cap/km^2. As it is surrounded by shallow waters, the island can be accessed by canoe and diesel-engine boats only during high tide. In light of this, the island has remained quite 'remote' and people still live relatively traditional lifestyles.

The Nicobarese follow an elaborate calendar of festivals and ceremonies that sometimes extend for months. Rituals and ceremonies are strongly linked to the regulation and use of natural resources, with celebrations ranging from commemorating the dead, to the welcome of the monsoon or the annual pig festival at the onset of the northeast winds, to name but a few. The organisation of such festivals can be time-consuming and resource-intense and requires a strong cooperation between family and community members.

Owing to their location on a historically significant sea route to Southeast Asia, the scattered Nicobar Islands have offered an attractive stopping place for trading vessels for a very long time. From the fifteenth to the nineteenth century, the islands were dominated by Indian, Arab and European trading ships offering a huge array of goods such as rice, cotton and timber to the Nicobarese in exchange for coconuts and other forest foods. The Nicobar Islands were officially subjugated to British domination in 1869. Trade was promoted and a system of general administration was set up that regulated trade through trading licenses. While coconuts formed the bulk of exported items, the cargo also included varied quantities of arecanuts, empty coconut shells (used in the manufacture of *hookah* to smoke tobacco), seashells, rattan, tortoise shells, sea-cucumbers and ambergris, the latter two regarded as valuables on the international market. Due to their long trading history they have enjoyed contact with the outside world for many centuries. Over time, sporadic barter trade came to be an integral part of the economic activities of the natives, even though it was of low commercial importance to the passing traders themselves. Iron, cloth and foreign foods in exchange for copra came to take a high symbolic value among the locals. Various traded items became embedded in the peoples' daily subsistence, and are today seen as symbols of wealth to be displayed at rituals and festivities. Imported rice became a staple and gradually replaced pandanus (a wild fruit rich in pulp and fibre) as the main source of carbohydrate. The former became much easier to obtain and prepare, compared to the dreary and time-consuming task of processing pandanus into an edible form.

Today, Trinket's economic portfolio combines fishing, hunting and gathering, pig and chicken rearing and the barter of copra (dehydrated coconut) in exchange for rice, cloth and other market commodities. Some families maintain small kitchen gardens that contain a variety of crops such as bananas, yams and jackfruit. The main source of nutrition, however, comes from seafood from the surrounding mangroves and from long gathering trips where they can pick and chose from a variety of edible leaves, fruits and tubers. Their major source of protein for the local population is obtained from the sea in the form of fish, shells, sea-cucumber and other marine food varieties. The people cultivate coconut plantations, which require only small labour inputs. Once planted, coconut trees can be harvested for about 100 years. Of the total coconut harvest from coconut plantations, half is processed into copra, a third is fed to pigs, and the remainder is consumed by chickens and humans. As ownership of the land is by joint family systems, each stretch of land has a distinct owner, or at least, clearly defined usufruct rights (Singh 2003a).

Since a couple of decades, the Indian government has introduced various welfare programmes to the Nicobar Islands. These include health services, education and transport infrastructure as well as cheap diesel and kerosene subsidies. The latter has instigated high fossil fuel consumption among the locals, mainly to run diesel boats to transport copra to and market goods from the market (see Singh 2003a).

6.3 Agricultural Intensification and Regime Transitions

The best known scholar in the field of agricultural intensification theory is probably the agricultural economist Ester Boserup (1965, 1981), as she developed her theory based on population growth, technical change and land and labour use intensification. By classifying successive land use intensities through increased cropping frequencies (forest-fallow, bush-fallow, short-fallow, annual and multi cropping), she inspired a dynamic land use model that had not existed before and verified it with a large number of empirical cases. To her, population growth was the prime mover for land use intensification. Another contributor to intensification theory was Wilkinson (1973), who extended the intensification theme 'backwards' to include hunters and gatherers and 'forward' to cover industrial societies (see Carlstein 1982: 11–12). To his mind, there exist two kinds of societies: those who live in ecological equilibrium and those where this balance is disturbed by population growth. For the former type of society, there is no need to apply innovative methods in terms of land use; for the latter, however, innovation in intensifying land use and improving technologies, have to be found.[6] In contrast to Boserup and Wilkinson, Sahlins (1972) proposed a different approach to intensification and contends the

[6] In his publication, Wilkinson (1973) gives an interesting account of how the Industrial Revolution in England was directly related to a shortage of land and land-based resources (e.g., wood), which, in turn, led to the exploitation of coal.

prime mover for the intensification of production to be solely due to political mobilisation within the kinship framework. The scope of societies he examined is limited and ranges from hunter and gatherers to agricultural systems that are only slightly more intensive than swidden systems (the so-called 'domestic mode of production'). What all theorists agree on, however, is that land use intensification results in higher work-loads and working hours.

So if this link between population growth, land use intensification and higher work-loads on individuals persists, a system would inexorably run into a deadlock. We know, however, that the introduction of fossil fuels, however unsustainable in the long run it may be, has provided a way out of this trap. And this is where the theory of sociometabolic regimes provides a useful framework that helps our understanding of contemporary sociometabolic regime transitions and the long-term environmental costs borne by such a transition.

The theory of sociometabolic transitions goes back to Sieferle (1997) and Weisz et al. (1999) who distinguish different sociometabolic regimes and believe energy limits to be the prime mover for societies to change, or, in Boserup's terms, to intensify land use. Their argument, based on historical analysis of hunter and gatherers, agrarian regimes and industrial societies, establishes a link between the availability of energy sources and sustainability. According to Sieferle, hunter and gatherers sustained themselves through a 'passive' use of solar energy, meaning that their metabolism relied solely on the existing density of solar radiation and its conversion into plant biomass within the territory they have access to. As they did not intentionally intervene in natural processes, they were essentially dependent on the natural rate of animal and plant reproduction; they could do in fact very little to raise the output per unit of land, however, they could easily decrease it (see also Harris 1977: 11). This is why the only sustainability threat for hunter and gatherer regimes was the depletion of those key resources necessary for their survival. The historical agrarian regime can be characterised by an 'active' profile of energy utilisation, as people controlled the rate of plant reproduction. They did so by intervening in the solar energy transformation process through the use of human and/or animal labour. People now cleared forests to cultivate agro-ecosystems, and were now engaged in stock breeding of those species useful for their subsistence. Within this regime, the resource density was manipulated to yield many more products useful for humans than would have been the case in the absence of constant human intervention. Still, the difficulty for these societies was with sustaining the fine balance between population growth and the agricultural labour force needed to maintain the productivity of the ecosystem and the preservation of soil fertility to ensure adequate yields (Netting 1993). The currently prevailing sociometabolic regime in many parts of the world is the industrial regime, which has largely overcome these limits through the introduction of a land-independent source of energy, namely the use of fossil fuels. It need not automatically be visible at the local level, since it may occur at higher scale levels and indirectly impact local agriculture (e.g., through the formation of transport systems). On the local level, therefore, one may see an intermediate stage between an agrarian (solar) and industrial (fossil fuel-based) regime in the form of indirect fossil fuel use. In such a way, the theory of sociometabolic transitions overrides Boserup's theory on decreasing labour and population density, since the increased use of fossil fuels and industrial technologies also lead to an increase in labour productivity.

Figure 6.1. is a matrix frame combining some of these theoretical notions.[7] The vertical axis lists the stages of market integration, from subsistence via sale of agricultural surplus to market production. As we are dealing with rural agricultural communities, I have tried to make a further subdivision for the latter two stages. These entail a distinction between independent and incorporated agriculture, both terms that were created by Bolhuis and Van der Ploeg (1985) in Hobbes (2005: 200) in order to designate a community's degree of involvement in (and hence dependency on) outside markets. In their view, independent agriculture comprises a high degree of involvement on the output side, however, without the need for external inputs. Incorporated agriculture, on the other hand, involves a high degree of market incorporation on the input side. By integrating these ideas, I hope to facilitate a

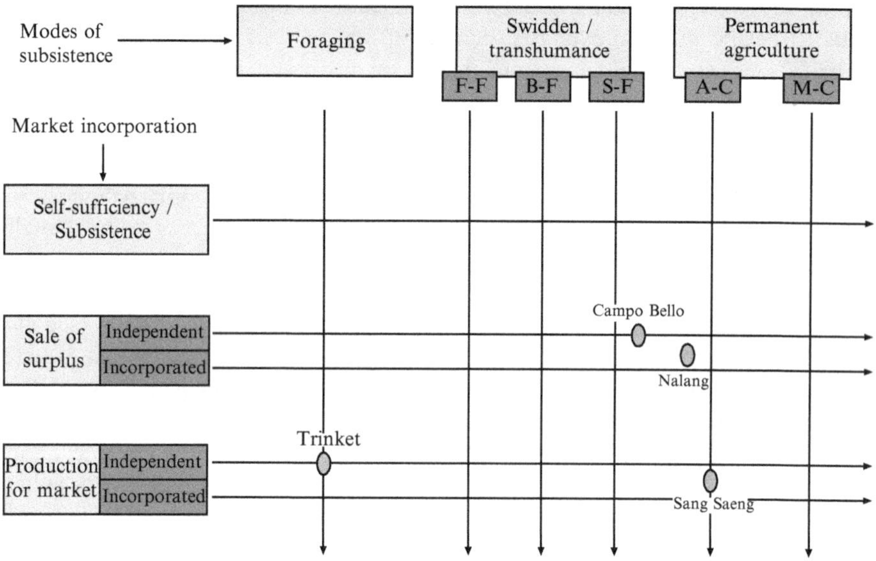

Fig. 6.1 Schematic representation of transitional stages

[7] This figure has been inspired by many scholars. The first idea came from Grünbühel et al. (2007: 149), who designed a scale based on the principle of a Guttman scale in order to measure their hypothesis of the existence of 'an in-built irreversible dynamics from one stage to the next'. Following their publication, many useful suggestions to enhance the logic and clarity of the figure were brought forward by Christian Lauk during a meeting at the Institute of Social Ecology in April 2009. His thoughts on separating the land use from the production practices, as was the case originally, inspired the structure of this figure. Moreover, I found Boserup's (1965: 15–16) grouping of five types of land use useful in order to obtain a more differentiated basis upon which to ground the four empirical cases. Lastly, Hobbes' (2005) article on material flow accounting in rural communities was instructive for making a subdivision between independent and incorporated (agriculture) within the stages 'sale of agricultural surplus' and 'production for the market'.

conceptual link between the local level and the market at higher scale levels, in which more rural communities are increasingly engaged. Returning to our schematic representation, the horizontal axis determines the modes of subsistence or land use practices, from extensive to more intensive forms of interacting with the environment. In order to obtain a more detailed account on the specific utilisation of land within 'swidden/transhumance' and 'permanent agriculture' classification, I have integrated Boserup's (1965: 15–16) grouping of five types of land use as mentioned above. Abbreviations stand for forest-fallow (F-F), bush-fallow (B-F), short-fallow, annual cropping (A-C) and multi cropping (M-C).

With the framework in place, the next step was to order the four cases sequentially according to their mode of subsistence and level of market incorporation. Let us look at Campo Bello first, whose land use practice I have placed near enough short-fallow cultivation, however, with a direction toward permanent cropping. This may sound peculiar for some, as swidden agriculture is indeed the prevalent land use system in the community. The answer for this decision has to do with how fallows are handled by the Tsimane'. Contrary to the ideal type of fallow – that a clearing is left uncultivated for a certain time period – the villagers of Campo Bello do not abandon fallows completely but rather cultivate them with species like palm trees or plantains. These sites are thus still part of the local production area, albeit on a much more extensive level. With regard to market penetration, things are much clearer. Crops are not planted especially for the market and only surplus produce (mainly rice and plantains) are traded in exchange for processed foods, soap and clothes. The system of Campo Bello does not rely on a high degree of external inputs for its agricultural production. Nalang's mode of production, on the other hand, constitutes permanent paddy rice cropping with a small fraction of shifting cultivation. Agricultural production is not geared to production for the market, yet entails the sales of agricultural surplus.[8] Contrary to Campo Bello, Nalang has already incorporated external agricultural inputs such as irrigation channels and motor-ploughs. SangSaeng's land use I have classified as permanent agriculture and, due to its paddy rice production, subdivided into annual cropping, since the paddy land is left uncultivated for several months between the rice harvest and next year's planting. Of all the four cases, SangSaeng displays the highest level of market incorporation. This assumption I base on the fact that the system cultivates a particular rice species that is entirely destined for the local market. I have placed the community in between independent and incorporated agriculture, as people have incorporated industrial elements such as chemical fertilisers or water pumps into their rice production system. Finally, the fourth case, Trinket, comprises a foraging society that combines fishing, foraging, pig rearing and the barter of copra. I have hence placed the system under the foraging mode, however, with a relatively high degree of

[8] Recently, some Nalang families have introduced the planting of cucumbers that are sold on the local market. It is the only major cash crop in the village besides bananas (Mayrhofer-Grünbühel 2004: 101-102). Notwithstanding, due to its only minor importance as compared to subsistence rice production (dry season cucumber cultivation accounts for 13% of all agricultural harvests), I have avoided to categorise the production modes of Nalang as 'production for market'.

market incorporation due to the frequent exchange of copra for rice or cloth. As copra is produced without any external inputs, it has been classified as 'independent'.

According to this taxonomy, we find Campo Bello, Nalang and SangSaeng along a continuum of growing land use intensification and market incorporation. Trinket, however, appears to deviate from this assumption as it has the most extensive subsistence mode of all the cases, but along with SangSaeng, produces especially for the market. But these impressions still need to be confirmed by a quantitative study to ascertain whether these cases show the metabolic profiles we would expect in light of the theories discussed above. I am hence going to explore a variety of comparative sets of biophysical indicators comprising ecological, agro-economic and labour time parameters; an analysis that will provide insights into the local biophysical changes taking place during a transition.

6.4 Biophysical Indicators of Intensification

6.4.1 Ecological Parameters

Examining the appropriation of energy and its efficiency provides a good starting point. To this end, I am going to look at the human-induced land cover changes (i.e., clearing of forest for agricultural and settlement areas and pastures or crop harvesting) by applying the indicator HANPP. Does the intensity of colonization activities indeed increase as systems move from one mode of subsistence to another? Also, another indicator of interest is the percentage of colonizing efficiency. Ideally, the most sustainable agricultural system would be the one that provides the maximum services to the social system (e.g., through continuous high yields) at the lowest environmental costs possible (see also Grünbühel et al. 2007: 153) (Table 6.1).

In line with theoretical assumptions, the most extensive system, Trinket, displays the lowest appropriation level of net primary production, accounting for only 22% of potential biomass energy,[9] and harvesting only 1%. Inherently linked to the low harvest is the low colonizing efficiency at 3%. In terms of land use, 55% are still covered by primary forest and 32% by grassland, while the remainder is made up of mangroves, beaches, housing areas and horticultural gardens. Coconut plantations, the most important direct and indirect (trade for rice) staple, account for the negligent land area of 0.7%.

Following Trinket on the land intensification continuum is Campo Bello, which shows a higher percentage of HANPP (39%). Colonization activities, however, are still moderately low, a typical feature of extensive shifting cultivation. The villagers

[9]There is still some doubt as to whether the interior grasslands on Trinket exist due to natural processes or are a result of previous human colonization by Danish invaders. What has remained certain, however, is that the colonizers left cattle on Trinket that continues to maintain the grasslands as such and prevent succession. For this analysis, it is assumed that the grasslands are man-made (see Singh 2003a).

Table 6.1 Colonization of land: HANPP, harvest and colonizing efficiency (source Grünbühel et al. 2007)

	Trinket	Campo Bello	Nalang	SangSaeng
NPP_0 (TJ·a)	1,090	244	640	36
HANPP (TJ·a)	234	94	343	28
HANPP (% of $NPP_{0)}$	22	39	54	77
Harvest (TJ·a)	7	9	24	11
Harvest (% of $NPP_{0)}$	1	5	4	30
Efficiency ($NPP_{harvest}$/HANPP)(%)	3	10	7	39

harvest 5% of the potentially available biomass energy, leaving sufficient NPP in the ecosystem to maintain its reproductive capacities. Despite regular slash-and-burn activities, the system's colonizing efficiency accounts for 10%. Concerning its land cover, the community still has access to some virgin forest resources, 47% of the entire village area is still covered by primary evergreen seasonal forest, an important hunting ground. Secondary forest accounts for 22% of the village area. Almost 10% are made up of savannah lands that are unsuitable for agricultural activity. The figures show that, at least for now, the sum of colonizing interventions is still fairly low. The agricultural land area accounts for a mere 11% of the total land resources within the community, which implies that there is still plenty of scope for extending agricultural production areas.

Interestingly, Nalang appropriates more HANPP (54%) than Campo Bello, but harvests less and has a lower colonizing efficiency than its counterpart. What this indicates is that the burning of forest cover among the local shifting cultivators (albeit marginal in area size) in Nalang destroys more biomass energy stored in the soil than in Campo Bello, where shifting cultivation based on the use of fire is the prevailing mode of subsistence. Be that as it may, Nalang still has large areas of grassland (10%) and secondary forest (31%) that show a high annual NPP production of 640 TJ due to the comparatively fast natural succession of secondary forest areas. Moreover, similar to Campo Bello, Nalang still contains large dense forest areas, covering about 43% of its entire territory. Adding to this, the area of (permanent) agricultural land use is even slightly smaller than in Campo Bello, accounting for 8%. It seems that for the time being, Nalang still has plenty of land area available for a possible extension of agricultural production in the future.

Finally SangSaeng, the system with the most intensive land use, shows the highest HANPP (77%) and the highest colonizing efficiency (39%) of all four cases. These numbers are invariably a result of the relatively intense use of chemical fertilisers and, but to a much lesser degree, fossil fuel-based applications for the production of rice. Adding to this, in the absence of swidden agriculture, SangSaeng can make permanent use of the agricultural area available to the system, thereby enabling the population to substantially increase biomass energy harvests. At the same time though, this land use strategy leads the system into a sustainability trap: SangSaeng uses almost 80% of its land for its paddy rice production, while, in the absence of extensive grassland areas, the remaining 18% comprises degraded forest areas (Grünbühel et al. 2007: 154). It is noteworthy that the system's high colonizing efficiency, which already depends to some extent on the use of external inputs, will only be sustained through a growing use of these inputs from the market to maintain the degree of soil fertility needed to produce continuous high yields.

If these four cases permit even a tentative conclusion, we can observe the following: in terms of the human appropriation of potential biomass energy, the intensity of intervention indeed increases as systems move from extensive to more intensive forms of land use. We also find that especially in shifting cultivation systems due to the use of fire, not all biomass energy appropriated by humans automatically leads to a higher colonizing efficiency. And, the case of SangSaeng illustrates that fossil fuelled inputs (e.g., chemical fertiliser) in agriculture can increase the colonizing efficiency of the system. Permanent agricultural systems no longer require the use of fire nor fallow periods for the soil to recover fertility. In the long-run, nonetheless, increasingly more of these external inputs are needed to merely maintain the same harvest outputs.

Next, I would like to examine two sets of agro-economic parameters that measure more precisely the level of agricultural efficiencies. These entail agricultural intensification indicators, on the one hand, and food security and dependency indicators, on the other hand.

6.4.2 Agro-economic Parameters

6.4.2.1 Agricultural Intensification

As we are dealing with agricultural systems, let us take a closer look at the biophysical dimension of intensification (Table 6.2). The intensification of agricultural land use in order to increase agricultural production per area can take place either through (1) higher material inputs or (2) higher labour inputs. Concerning the former, we can examine fossil fuel inputs and the use of imported agricultural machines. In our cases, fossil fuel inputs appear to augment with land use intensi-

Table 6.2 Agricultural intensification indicators (source Fischer-Kowalski et al. forthcoming)

	Trinket	Campo Bello	Nalang	SangSaeng
Direct Material Input (DMI) (t/cap·a)	2.4	1.9	3.0	4.4
Direct Material Consumption (DMC) (t/cap·a)	2.3	1.6	2.6	3.2
Fossil fuel inputs (GJ/ha)	0.0 (copra) 55.0 (rice)	0.0	0.7	2.2
Labour time per area (h/ha·a)	460.0 (copra) 527.0 (rice)	549.0	1,534.0	570.0
Labour productivity (MJ/h)	237.0 (copra) 300.0 (rice)	14.0	13.0	38.0

fication for food production. This is clearly visible with Campo Bello, Nalang and SangSaeng, as they range from no inputs in the most extensive agricultural system of Campo Bello to 0.7 GJ/ha in Nalang and 2.2 GJ/ha in the most industrial agricultural system of the three cases, SangSaeng. As for Trinket, fossil fuel inputs into food production are more indirect. Whereas no fossil fuel inputs are required for the production of copra, things are different for rice. Rice is traded on the market in exchange for copra and state-subsidised fossil fuel inputs are required for the motorboats that transport goods to and from the market. As far as the use of agricultural tools and machines is concerned, Boserup (1965: 27) distinguishes between the kind of tools and the making of tools and points to three basic types of agricultural change: (1) the change from one kind of tool to another, e.g., from digging stick to hoe, (2) the change from the make of tools from home-made to factory-made, and (3) the change to better makes of tools but without changing the kind of tool. Her distinction is especially applicable to Campo Bello, where some families still use a dibble stick for the cultivation of rice. Others apply a different kind of tool, namely rice seeders. Also, most families own factory-made hoes, spades or axes, while a small number of farmers still use crude iron axes. We see that while people can choose between the efficiency of makes of particular kinds of tools, these technical changes are not necessarily linked to changes in the land use system. With respect to Nalang and SangSaeng, both use fossil-fuelled agricultural inputs which we see reflected in the fossil fuel input indicator. Nalang, on the one hand, uses motor-ploughs (22) that serve both as an agricultural tool and as a transport vehicle. These 'iron buffaloes' as they are referred to locally, have largely replaced traditional buffaloes (Mayrhofer-Grünbühel 2004: 131). While only indirectly related to agricultural production, other fossil-fuelled vehicles include five motorcycles and one pick-up truck. Finally, SangSaeng displays the most intense use of agricultural machinery that entails the application of ploughs (44), motor-ploughs (2) and electrical water pumps (6). Adding to these are pick-up trucks (8), motorcycles (20) and other vehicles that serve for the transport of agricultural produce. Interestingly, the use of motor-ploughs is lower in SangSaeng as compared to Nalang. On the whole, however, (external) direct and indirect material inputs in food production are highest in SangSaeng. In part, this is supported by the indicator direct material inputs (DMI), which is lowest in Campo Bello (1.9 t/cap·a), higher in Nalang (3.0 t/cap·a) and highest in SangSaneg (4.4 t/cap·a). As to Trinket, the comparatively high direct material inputs of 2.4 t/cap·a are largely due to the extraction of sand and the import of cement, both of which are considerably used as local construction material as well as exported to other islands.

The second way of intensifying agricultural production is through a higher labour input per hectare. Labour inputs in food production are highest in Nalang, where production is more intensive than in Campo Bello or Trinket. So far, the Boserup thesis that labour time increases with more intense land can be supported empirically. As compared to the more extensive swidden agriculture of Campo Bello, Nalang's permanent agriculture with only marginal shifting cultivation demands substantially more labour inputs such as the flooding of paddy fields or transplanting. Campo Bello's dry rice and other staple food productions, on the other hand, can be done without these extra activities. If we look at SangSaeng, however, the much lower labour inputs required per area run counter to intensification theory. Here, sociometabolic transition theory may provide a possible explanation. In this respect, it is the change in the energy systems towards a more fossil fuel intensive agricultural production, which is responsible for the significant reduction of human labour inputs.

One relief to the work-load of humans in agricultural production is the use of animal labour. In both the Nalang and SangSaeng systems, buffaloes are used during the rice production cycle. We know that in SangSaeng, the use of buffaloes as draught animals is restricted to the preparation of rice fields, contributing a mere 4% of total labour investments (Grünbühel et al. 1999). In Nalang, with the advent of engine-powered ploughs, buffaloes have lost their former importance as working animals for ploughing of rice fields and, due to their easier maintenance and higher reproduction rates, cattle has been gradually replacing buffaloes Mayrhofer-Grünbühel (2004: 113). As to Campo Bello, the food production system does not comprise animal labour and it is likely to remain this way. First, there is no habit for animal (cattle) labour in the region. Second, a recently opened Tsimane' field contains various tree stumps dotted all over the field area; a fact that impedes cattle to move along smoothly. Third, the Tsimane' in the area do not have much experience in cattle rearing. Fourth, if we remember their relatively low agricultural labour time, the people of Campo Bello do not feel an immediate need to increase their output per labour-hour that would be obtained by the introduction of animal labour. Finally, let us examine the productivity of labour. Here, the highest productivity rates are found with Trinket, which also manifests the lowest labour requirements for food production. Nalang and Campo Bello, on the other hand, reveal much lower labour productivities, whereas SangSaeng's productivity gains are on the upper end of the scale. Intensification theory alone (a decline in individual productivity leads to people working longer hours), does not suffice to explain our cases. Again, we may refer to the theory of sociometabolic regimes, as it establishes a clear link between an increase of labour productivity and the use of fossil fuels in the production system; which is the case with SangSaeng.

6.4.2.2 Food Security, Self-Sufficiency and Dependency

Another set of biophysical indictors considers the degree of self-sufficiency and involvement in external markets. To this end, I will first of all examine food security levels in each of our communities and then look at what role markets play in their respective diets. Measuring market dependency will be achieved by the percentage of food imports relative to consumption, on the one hand, and food exports relative to food production, on the other hand.

For a definition of food security I would like to resort to Hobbes' (2005: 201), who describes it as 'the degree to which one can grow, extract or buy the calories one needs'. This definition designates food security as an independent variable of food abundance and people do not necessarily have to grow their own food to be food secure. This, to my mind, is a good starting point for an analysis of the different strategies undertaken by each of our cases to ensure food security. As we can see, all four cases show adequate daily food intakes, at least in calorific terms.[10] But what about the different nutritional energy sources the systems draw from? To start with Trinket, we see that the system derives relatively little nutritional energy from domestic agriculture and imports, a mere 31%. This is not necessarily surprising, since colonization activities are restricted to the cultivation of kitchen gardens and extensive coconut plantations (the latter has even been described as a form of gathering, Simron Jit Singh: pers. commun.). Befitting their foraging mode, the large remainder is extracted from foraging and fishing, accounting for 69%. At first glance, this extensive production mode seems ideal. The inhabitants of Trinket have a varied diet and do not have to invest much of their time and energy resources in agricultural production. Their staple, rice (although not eaten universally) is obtained via the exchange of copra; which requires much less intensive forms of land use than would be needed for the production of rice. On the downside, however, this trading mechanism bears the cost of high fossil fuel consumption and remains favourable only as long as the fossil fuel subsidies given by the Indian government remain in place. What is noteworthy is the fact that, not so long ago, the people of Trinket used to produce their own staple, pandanus, but have gradually replaced it with imported rice over the last century.

In contrast to Trinket, Campo Bello extracts substantially more nutritional energy from agricultural production that, including a small fraction of food imports, accounts for 84% of the community's nutritional intakes (Table 6.3). Food security

Table 6.3 Food security and dependency indicators (source: Grünbühel et al. 2007; Fischer-Kowalski et al. forthcoming)

	Trinket	Campo Bello	Nalang	SangSaeng
Daily food intakes (Kcal/cap·d)	2,876	2,664	2,681	3,215
Rice consumption (Kg/cap·a)	64	45	277	233
Percentage of staple locally produced (%)	0	100	100	100
Nutritional energy from agriculture, incl. imports (%)	31	84	89	90
Nutritional energy from hunting, fishing, gathering (%)	69	16	11	10
Food imports relative to consumption (% in GJ)	31	20	7	42
Food exports relative to food production (% in GJ)	100 (copra) 0 (rice)	37.5	17.4	82.5
Exported crops (%)	49	34	18	52
Exported rice (%)	0	52	11	55
Exported Livestock (%)	0	11	5	30

[10] According to FAO data, recommended minimum daily intakes are 2,500 kcal/d – FAO (2001).

does not seem under threat since bananas, manioc and all kinds of fruits and vegetables can be harvested throughout the year. Adding to this, there are no major food taboos attached to these food sources. Rice, on the other hand, is harvested once a year only and families repeatedly sell a large fraction of their entire production immediately after harvest, often in exchange for pasta noodles. It seems that imported pasta noodles are on the way towards replacing rice as the main staple. In 2004, per capita rice consumption amounted to 45 kg, compared to 36 kg for pasta noodles. Similar to the staple food transition that took place in Trinket some time back, it seems that Campo Bello is currently experiencing a gradual replacement of the source of carbohydrates from a traditional crop to an imported food staple. There is a difference, however: the people in Campo Bello do not change their land use and production in favour of more extensive and less labour-extensive modes, but retain dry rice farming. This strategy, coupled with hunting, gathering and fishing (16%) has the advantage that people can chose between the direct consumption of rice and the exchange of rice for pasta noodles. Both options are valid within the above definition of food security. There appears yet another distinctive feature between these two food transitions, despite lacking more precise historical data. Whilst it seems that Trinket's transition from pandanus to imported staple foods was driven by exogenous factors (i.e., state subsidies), Campo Bello residents appear to draw from the market those processed foodstuffs they consider a delicious supplement to their daily diet. Influenced by the changing status of imported foods among the Tsimane' population, the people of Campo Bello have also been adjusting their taste. In this respect, they have not been driven by state interventions 'pushing' them to the market.

Nalang shows a similar calorific profile to the former two cases; however, here the consumption of rice is substantially more important. Rice is consumed three times daily and provides a small surplus that is stored until the next harvest. The local environment also offers a great variety of edible plants and animals, whereby hunting, gathering fishing returns provide a welcomed variation to the local rice-based diet (11% of all nutritional energy extracted and imported). This, coupled with the fact that population density is low, as I discuss below in more detail, and the fact that no major food taboos exist makes Nalang a food secure place.

The highest amount of nutritional energy deriving from agriculture and imports can be observed in SangSaeng, while the proportion of nutritional energy derived from foraging activities is lowest (10%). This correlates with the fact that SangSaeng is the most market-integrated and most land use intensive of all the cases. Since money is available due to cash crop rice production and a large number of migrant workers, people can buy processed noodles, fish sauce and meat from the market (Grünbühel et al. 2007: 157). Similar to Trinket, SangSaeng's food security is thus partially based on the purchase of calories rather than domestic extraction.[11]

[11] The means of market exchange, however, differs between SangSaeng and Trinket. Whereas SangSaeng's market transactions are done via cash, Trinket's exchange mechanisms are largely based on copra barter trade.

This leads us to quantitatively examine the role markets play in food provision in each of our four cases. If we take the degree of imports and exports as a measure of market integration, then Nalang stands out as the most independent and self-contained of all. The market adds a mere 7% (in GJ) to the system's food consumption, and exports are, though twice as high as their imports, still lowest compared to the other cases. Almost three-quarters of all agricultural production are in fact consumed within the village. Rice is traditionally stored until shortly before the following harvest and marketed only after internal household's needs have been met. We see that the village follows a sustainable smallholder strategy (see Netting 1993) that focuses on meeting the nutritional needs of its inhabitants rather than engaging in deliberate market production. Having said this, however, the agricultural system may not be called subsistent in the narrowest sense if we recall Bolhuis and Van der Ploeg's (1985) classification of agricultural types, as the people of Nalang depend on fuel inputs for the motor-ploughs as well as other tools from the market they use for the production of rice. In this respect, we find the agricultural system of Nalang to resemble more the 'incorporated' type. Adding to this, due to the growing need for money (e.g., to pay taxes and agricultural inputs, for example), the system has reacted by diversifying their production base to add non-traditional cash crops like cucumbers rather than intensifying the traditional production of rice. After rice harvests, abandoned fields are turned into cucumber gardens and crops raised are entirely sold at markets close by. Also, but on a much lower scale, families have expanded their traditional banana gardens in order to market the surplus. Both crops can be transported without any difficulties and endure several days without spoilage. In sum, we can say that the people of Nalang have reacted to increasing monetary pressures through a mix of diversification (cucumbers) and intensification strategies (bananas), while at the same time trying to maintain their rather self-contained and sustainable staple production and consumption of rice.

A closer look at Campo Bello's import/export profile reveals a different picture. Food imports are almost three times higher than those in Nalang, a fact that can partly be explained by the high energy food imports like sugar or pasta noodles to Campo Bello. On the export side we find similar results. Campo Bello's exports account for more than twice as much as in Nalang, and rice makes up for the bulk of exports. Not only are more than half of all rice harvests exported, but the timing of export also differs between the two cases: while we see a fairly even distribution of rice sales throughout the year in Nalang, the people of Campo Bello are more inclined to sell large quantities shortly after rice harvests. As the sale of plantains, manioc and maize complement the export profile of Campo Bello, we see that the system has opted for intensifying the production of staples rather than deciding on the production of non-traditional cash crops. Notwithstanding, if we shift the focus from a mere food import/export analysis to scrutinising the types of market inputs needed for food production, things are somewhat different. For the time being in Campo Bello, the need for technologies to produce the staples is limited and fossil fuels are not yet part of agricultural food production. Moreover, the increasing demand for cash primarily belongs to other necessities (e.g., school materials) or desires rather than investments in the food production system (there is no need to pay for land taxes, for example). In such a way, the type of agricultural production practiced in Campo Bello, that is high exports without relying on many external inputs, may be referred to as 'independent agriculture' (see Bolhuis and Van der Ploeg 1985). This entails a much lower degree of market integration and cultural impact as would be the case with incorporation on the input side (Hobbes 2005).

SangSaeng follows yet another strategy of connecting to the market. This entails the intensification of the main staple crop, rice; but by diversifying its variety, rather than investing in additional dry-season production. This is a result of the ecological constraints the community is facing, since land resources are scarce and fertile farmland is no longer available to the villagers. Similarly, the region's dry climate prevents any dry-season farming. At the same time, since the type of glutinous rice produced locally does not suit market demands, local producers had to opt for the production of higher-quality white Jasmine rice; 55% of total rice production is destined for the market and not consumed by the local population. It goes not without saying, however, that the cash crop varieties planted within the community are subsidised by the local administration, as they all require additional chemical fertilising. Moreover, the export of livestock has gained importance in recent years, accounting for 30% of the total animal stock. Besides their significance as working animals, they are also treasured as a source of wealth and in times of need they can easily be sold at cattle markets. In terms of the import/export ratio relative to food consumption and production, the community of SangSaeng is undeniably the most market-integrated system as compared to the previous two cases. On the input side, food imports are six times higher than those in Nalang and twice the amount of Campo Bello. This is not surprising if we take into account the cultural changes that have taken place through the experiences of migrant workers to urban centres. Marketing campaigns, at the same time, and the unremitting influence of mass media have changed values within the social fabric and strengthened the importance of foreign foods as status symbols. On the output side, food exports even double the amount of imports, a number that reflects quite well the production of one product, namely Jasmine rice, exclusively for the market. Market integration has further been fuelled by the extensive network of paved roads in the region, that makes the transport of commodities reasonably inexpensive and time-efficient.

Trinket presents an interesting case as it clearly demonstrates that market integration is not necessarily linked to the intensification of agricultural production and land use. What we see is that Trinket is engaged in the regional market through the sale of one product: dehydrated coconuts in the form of copra. It is somewhat used as a currency to obtain some of the food staples (e.g., rice) as well as other everyday commodities. One third of all food consumed on the island is in fact imported, while the entire copra production is destined for the market. However, fresh coconuts also serve subsistence and account for more than half of the entire coconut production. At the same time, when people feel the need for copra, they simply harvest more coconuts and transform them into copra. Except for copra, no other domestic products from gardening, fishing and gathering ever leave the island boundaries. The export of livestock is equally non-existent. Pigs have a strong religious meaning and are needed for the celebration of ceremonies or rituals.

6.4.3 Human Labour Time Parameters

The last set of parameters I am going to look at is the use of human labour time. In Chapter 5 I have looked at the use of human labour from a generic angle, trying to understand the dynamics of human time use across different age/sex categories within the system of Campo Bello. What I am now interested in is the comparison of human labour time in different world settings. In line with intensification theorists, the hypothesis that needs testing is the assertion that labour time increases when systems move from more extensive to more intensive modes of subsistence. For this comparative endeavour only data from Trinket, Campo Bello and Nalang can be drawn from. Regarding SangSaeng, although first efforts to record time were made, the data sets generated cannot be presented in a comparable way to the other three cases. At the end of this section, I will nonetheless add information on SangSaeng's labour profile, albeit on a qualitative note only (Table 6.4).

A first look at the economic system reveals that the data seems to correlate with intensification hypothesis. Trinket's time inputs into the economic system comprise only a quarter of the time invested in Campo Bello and one fifth of the time inputs in Nalang. What we see indeed is a substantial increase in labour time from extensive to more intensive forms of land use. This is especially reflected in the agricultural and horticultural activity set, where an average adult person in Nalang, the system with the most intensive land use, invests about an hour more per day than in Campo Bello. This increase in labour can be traced back to the higher number of labour tasks needed for the production of paddy rice as compared to dry-land rice. On the other hand, the cultivation of cucumbers also requires some indirect labour inputs such as the installation of wooden poles for supporting the plants, and fences that impede the invasion of livestock into the garden area. Direct labour efforts, on the other hand, are limited and involve sporadic weeding, the application of small quantities of fertiliser and a gradual harvest period from the end of March until late April (Mayrhofer-Grünbühel 2004: 117). Banana gardens, on the other hand, simply require periodic weeding and the occasional clipping of dead leaves and branches. Campo Bello comes next with a lower labour input in agricultural activities, correlating well with its more extensive swidden practices. Moreover, staple food production is more varied and includes the much less labour-intensive growth of

Table 6.4 Comparison of labour time for the average adult 16–60 (source Fischer-Kowalski et al. forthcoming)

	Trinket (h/d)	Campo Bello (h/d)	Nalang (h/d)
Agriculture/horticulture	0.09	2.53	3.06
Animal husbandry	0.14	0.02	0.61
Hunting, fishing, gathering	0.58	0.98	0.59
Trading	0.39	0.43	0.00
Wage work	0.00	0.27	1.46
Handicraft	0.00	0.41	0.13
Total	1.20	4.64	5.85

maize, plantains and manioc.[12] Trinket's colonizing activities are limited to the maintenance of coconut plantations and small-scale kitchen gardens, both of which require little labour. Not surprisingly, labour investment in animal labour is highest in Nalang, the most permanent agricultural system of all three cases and the only one that uses animal labour for its staple food production. The extremely low labour inputs by the average adult in Campo Bello not only reflect the limited livestock numbers in the system, but also the fact that these tasks can liberally be delegated to children. If we look at Trinket, however, we find higher numbers, since the rearing of pigs requires comparatively more human time and effort. Hunting, fishing and gathering are important leisurely pastimes in Campo Bello. As Trinket is surrounded by the open sea, fishing constitutes a vital part of everyday life. In Nalang, the time expended on these activities is mostly taken up with fishing and the gathering of bamboo shoots and all kinds of edible forest fruits.

The remainder, that is trading, wage work and the manufacture of handicraft are inherently connected to the direct generation of income. Here we find that Campo Bello follows the most diversified strategy in terms of labour inputs, investing more or less the same amount of time resources to trading and the manufacture of handicrafts. Trading, however, also involves barter trade to some extent, which is not geared to monetary returns. Relatively little labour is invested in wage work, but this is hardly surprising if we take into account that only men were engaged in wage work at the time of data collection. Wage labour in Nalang, on the other hand, accounts for the lion's share of monetary returns, with the fabrication of saleable handicraft comprising a minor top-up only. Finally, Trinket's cash producing economic profile is solely based on the trading of copra, making it the most vulnerable of all three cases.

Being the most land use intensive of all the cases, we may expect SangSaeng to also have a much tighter labour time regime. And indeed, an average work-day in agriculture has been calculated at 6.7 h for an average adult, this almost equals the amount of an average work-day in industrial societies (see Fischer-Kowalski et al. forthcoming). From qualitative field reports we also know that agricultural workloads are unevenly distributed along the rice production cycle and the rice harvest period is dreaded by everyone exactly for the substantial work-loads ahead. Though lacking quantitative data on labour time invested in wage work, there is reason to assume that time investments are considerably high, since about one fifth of the village population frequently engages in temporary or permanent out-migration, especially during the less agricultural labour-intensive dry season. These labour migrants relieve the pressure on the already scarce local resource base as well as enable a substantial inflow of industrially produced goods into the community, either directly or indirectly in the form of remittances.

---

[12] Root crops especially are known for their low labour requirements. In Descola's (1996: 322) view, this is the reason why in the last four or five millennia Amazonian societies always planted manioc; an activity that did not engender less leisure time that could be used in hunting, gathering and fishing.

6.5 Driving Forces for Change

Having scrutinised a variety of useful indicators that help improve our biophysical understanding of different transition stages, we are still left with the pressing concern of what actually drives a sociometabolic regime to change. Changes of any kind may be driven by endogenous or exogenous forces. Intensification theorists largely claim endogenous processes, in the form of population growth and resource scarcity, to be the prime mover for agricultural systems to change. Does their theory, based largely on historical pre-industrial societies, or at least certain aspects of it, still prove useful for contemporary systems that are all, yet to different extents, part of a larger, exogenous world setting? And how revealing can the theory of sociometabolic transitions be in this context? This is why I find Grünbühel et al.'s (2007) examination of two exogenous processes, namely access to markets and state intervention, a useful proposition to follow.

6.5.1 Population Dynamics and Density

Let us first examine Boserup's (1965) central proposition that changes in land use and technological improvements occur from within agricultural communities, rather than outside interventions. Being the most 'advanced' in the agricultural production system, SangSaeng's population density is by far the highest, followed by Nalang and Campo Bello with a figure of less than half. Trinket has the lowest population density, 11 cap/km²; a figure that is still though much higher than that of a purely foraging society.[13] Nalang's and Campo Bello's population density profile correlates with estimates for agrarian societies, ranging somewhere between 30 and 40 cap/km² (see Krausmann and Haberl 2002), with Nalang being situated at the upper end of the density scale (Table 6.5).

SangSaeng's annual population growth rate is fairly moderate, at 1.5%. However, while more human labour would be needed to cover the shortages at the peak rice harvest period, the villages' drained local resources (degraded forest and soils), on

Table 6.5 Sociometabolic indicators on population, territory and artefacts (source: Grünbühel et al. 2007; Fischer-Kowalski et al. forthcoming)

	Trinket	Campo Bello	Nalang	SangSaeng
Population size	399	231	702	171
Size of territory	3,626	615	1,63	184
Population density (cap/km²)	11	38	43	93
Population growth (%)	1.5	3.8	3	1.4
Weight of artefacts (t/cap)	8.3	1.4	1.7	18.6

[13] Lee and De Vore (1968), for example, forward the figure 0.6 cap/km² for pure hunting and gathering societies of the Pleistocene era.

the one hand, coupled with its limited size, on the other hand, would simply not sustain a larger population. Well aware of these constraints, the coping mechanisms of many residents is temporal labour migration elsewhere. Since many of the migrants return for peak rice harvest times only, their absence during much of the year releases the overall pressure all together on the already exhausted local resource base. Yet this is no novel development as Isan society, according to Fukui (1993) in Mayrhofer-Grünbühel 2004: 158) has long engaged in temporary migratory processes as a function to control population growth and sustain an even population/environment balance. What needs to be taken into account is that permanent migration, on the other hand, would simply have to gradually increase in order to maintain a stable population balance within the community.

In Nalang we observe an annual population growth rate that is twice as high, at 3%. Contrary to SangSaeng, however, we find a resource base that is by far less degraded and a population density that is even less than half as high. But what does the future hold? Taking into account that Nalang's cultivable land areas for permanent agriculture are limited, the (Boserupian) strategy would be the intensification of land use (and hence labour inputs) on the existing agricultural areas in order to ensure feeding of the growing population. According to Grünbühel et al. (2007: 161), it is most likely that this process will lead to a qualitative transition of the local production system in the near future. One strategy would be the investment into irrigation schemes to boost the yields of dry-season cucumber cultivation. The need for fertiliser will increase in order to halt soil degradation and transports by fossil-fuelled mini-tractor to the markets of Bane NoneHai or MuangFuang will have to be organised more frequently. For the time being, there should be enough labour reserve for this strategy, but with an annual population growth that continues at 3%, the community is likely to run into a sustainability trap and outgrow the local resource base available to the villagers.[14] Equal to SangSaeng, increased out-migration will be the only coping strategy to preserve a 'healthy' balance between population and the local environment.

Campo Bello's annual population growth is the highest of all the cases, accounting for 3.8%. Whereas today the community is still endowed with large forest resources coupled with a fairly low population density, things are likely to change in the (not too distant) future. The people of Campo Bello will be faced with ever more diminishing game and fish resources, forcing them to relocate their labour resources to agricultural production. For the time being, potential agricultural land resources are still widely available; almost half of the village area still contains primary forest and roughly a quarter is still covered with secondary forests. Hence, people are likely to first extend the sizes of their rice fields rather than intensifying production. Extension strategies are also likely since land is a free resource and can freely be 'colonized'. Moreover, the village is endowed with a large enough labour

[14]At the present growth rate, Nalang will reach SangSaeng's population density within 27 years. It appears likely that Nalang will not be able to sustain such a growth rate for long, since land suitable for agriculture has already become scarce and the strategy of expansion has more or less reached its limit (Grünbühel et al. 2007: 175).

reserve to cover increasing labour requirements. An intensification of land use, on the other hand, will only occur at moderate levels (e.g., through the application of fertiliser). A complete shift from swidden agriculture to permanent farming, on the other hand, is very unlikely to take place. This has mainly to do with the relatively poor quality of local soils; they are leached and too acidic to sustain high-yielding permanent agriculture beyond the first and the second year of production. Opening new forest patches and burning the forest cover therefore provides a vital strategy that kills pathogens and releases the nutrients, which are stored in the forest biomass, into the soil (see, for example, Sanchez 1976). Today, people do not necessarily plant rice fields on alluvial river terraces, where soils are most fertile, yet opt for more remote, but secure sites.[15]

Can we expect Campo Bello's population to grow further in the not so distant future? The answer is yes. First, at the time of research, the village consisted of a large number of young families who are likely to expand their family size in the near future. Also, permanent out-migration is still not a common strategy in the community, and, while on the increase, a relatively strong aversion to engage in often exploitative work relationships with outsiders is still fairly prevalent within the community. Adding to this, if families expand their production of rice in order to buy processed foods and other commodities from the market, then the need for children to provide more labour will become ever more pronounced. Moreover, upon my last return to the community in 2008, population numbers had reached 250, as compared to 235 in 2006 and 231 in 2004/5. Finally, if we compare Campo Bello's annual growth rate with that of the whole Tsimane' ethnicity in the territory, which was calculated at 4.86% (Reyes-García 2001), we see that there is still room for 'increase'.

Finally, Trinket's annual population growth rate at 1.5%, albeit rather large by foraging standards, does not pose a threat to its local resource base. Adding to large stretches of land resources, the island is as surrounded by mangroves and the ocean, both rich sources of protein for local consumption. As the trading of coconuts in exchange for staple food and commodities is a year-round activity, the local population can thus avoid potential seasonal food shortages. Compared to the previous cases described, the people of Trinket need to invest very little labour time to meet subsistence needs and to produce for external markets.[16] Since Trinket itself was populated no earlier than 150 years ago, its population density today is less than that of the other islands in the archipelago with an average of 23 cap/km^2. Following this development path, Trinket is in the fortunate position to even double its population density, still without facing any major resource constraints. The local inhabitants enjoy yet another advantage. Given that some islands of the Nicobar archipelago are still uninhabited today, population pressure can be kept under control by an

[15] In January 2006, for example, river floods destroyed large amounts of rice and plantains that were cultivated on alluvial river terraces. Local people were suddenly deprived of their subsistence and economic base and had to resort to emergency food aid.

[16] An average of 14h/cap·a are invested in copra production; 121 h/cap·a are spent on fishing activities (Grünbühel et al. 2007: 172).

expansion strategy to these islands. Again, this correlates with Boserup's notion that the strategy of expansion, for as long as possible, precedes intensification. If we are to believe her ideas that productivity of labour time declines with land use intensification, then we can expect people to increase their labour time inputs merely to maintain their standards of living and outputs.

6.5.2 Access to Markets

The second driver to be assessed is access to markets. To what extent have market opportunities had an impact on local socio-economic subsistence structures?[17] And, are these changes reflected in their sociometabolic profiles?

Situated along a historically important sea route, Trinket has long been – in this case for centuries – exposed to outside markets, without affecting the basic structure of the people's modes of subsistence. Nonetheless, barter trade came to be an integral part of the local people's socio-economic fabric and became essential to their survival. Over the last 250 years, the inhabitants of Trinket have experienced an ever increasing integration into the regional and to some extent the global economy, insofar as a drop in global copra prices has had a direct impact on the local economic system. Since the mid-1980s, the copra industry has twice faced crisis, the first taking place in 1987 and the second in 2000. The government strategy to prevent the impoverishment of Nicobarese traders was the set up of the Copra Price Support Scheme, which guarantees subsidies for the copra price in an attempt to make the local population less vulnerable to world market fluctuations (Grünbühel et al. 2007: 163).

In the case of SangSaeng, market access is a much more recent phenomenon which began in the 1960s, when major national efforts were made to integrate the historically remote and isolated region of northeast Thailand into the national socio-economic fabric. Infrastructure developments (roads, rail, electricity and mass media) have since brought markets into the vicinity of formerly isolated communities and created new market opportunities via social networks and facilitated access to information. From a theoretical viewpoint, these development efforts have also brought about biophysical changes. Infrastructural changes have introduced an indirect energy (fossil fuel) subsidy to these previously solar energy-based rural communities, without which the local population would not have been able to join markets. For SangSaeng, this energy subsidy has allowed 'a way out of what would have become a deadlock involving resource over-utilization and finally pauperisation' (Grünbühel et al. 2007: 168). So it seems that the village was left with no other option but to join the market economy. Since land resources had become progressively scarcer, even moderate population growth would have placed a burden on the people's subsistence economy. Adding to these aspects was the deficiency of other essential resources

[17] In his book 'The Great Transformation' Polanyi (1957) contends that market exposure leads to a substantial change in a society's socio-cultural and subsistence structure.

(insufficient labour resources, seriously degraded forest resources, inadequate climatic conditions) to guarantee year-round subsistent food production.

SangSaeng's incorporation into the market has been gradual and started with a switch to higher quality cash crop rice in addition to the production of locally consumed rice. This more intense production for outside markets has in turn increased the pressure on land and labour resources. To generate more cash in order to make ends meets (need for fertiliser, taxes, and processed foods), the younger village population especially (almost 20%) now participates in the temporary outside labour market in the dry season. Finally, we can observe permanent migration to other communities or commercial hubs. Migratory flows, however, can only be maintained as long as the transport system remains in place, connecting the local system with the urban labour market.

Nalang presents a different scenario since the existence of markets has influenced the local production system to a much lesser extent. Contrary to SangSaeng, Nalang is located quite remotely in a valley without major transport facilities, some 180 km away from the regional market town. Market integration was fuelled by the construction of a road in 1980 and has since mostly taken place through the import of processed food items (e.g., alcohol, soft drinks, sugar) and various consumer goods (e.g., radio, bricks, motorised ploughs). Interestingly, the introduction of agricultural technology in the form of the motor plough, a consequence of government policies, has rather contributed to intensifying subsistence agriculture in response to increasing population densities instead of fuelling market integration. As we have learned previously, the major part of agricultural harvests goes into the subsistence of the system and only a small portion is exported via the market.

Likewise, Campo Bello's remote setting has largely contributed to a fairly low market integration level. This isolation does not necessarily imply a far geographical distance from the market, as a mere 24 km, or a days journey by canoe, separate the community from the market town of San Borja. It has more to do with general accessibility, as in the absence of a road, transport during the wet season is largely restricted to the river. Access to the market, particularly at rice harvest peaks, is provided via travelling river traders who visit the community. So we find that at certain times of the year, the market comes to the people of Campo Bello. In exchange for rice and plantains, the villagers obtain processed foodstuffs and all kinds of industrially produced items with a predominantly utilitarian function. Rifles and synthetic fishing nets, for example, are purchased to facilitate the subsistent activities the Tsimane' delightfully engage in. Here we see a parallel to Nalang, where many items obtained via the market also tend to contribute to an intensification of subsistent activities rather than increase production for the market. The impression I gathered from field surveys is that all families, albeit to differing degrees, make deliberate use of the market for their specific needs and wants rather than being pushed into the market economy. This notion somewhat coincides with the findings of an interesting longitudinal study undertaken by Godoy et al. (2005) on the determinants for market participation of the Tsimane' ethnic group. They found that the level of market integration, though higher in settlements closer to the commercial hub of San Borja, is inherently connected to conscious individual and household decisions.

If I am now to distinguish between 'push' and 'pull' determinants to market access and participation, I suggest to group together Trinket and SangSaeng, on the one hand, and Nalang and Campo Bello, on the other. As far the first grouping goes, I find that both systems have been somewhat 'pushed' into the market through the provision of direct and indirect state interventions and subsidies, and, in the case of SangSaeng at least, due to the scarcity of an adequate locally produced food base (except for rice). In SangSaeng, accessibility to transport infrastructure and communication has indirectly facilitated market access, whereas the provision of chemical fertilisers, hybrid rice seeds and loans has triggered more directly the exclusive production for the market. State subsidies play a major role in Trinket's market participation, as the government controls the exportation of copra at subsidised prices. If we look at the timing of market integration, we find that Trinket has had access to the world market for centuries, while SangSaeng's market integration was achieved by a deliberate regional development strategy starting in the 1960s.

As far as the second group is concerned, I find that Nalang and Campo Bello have both joined the market on a rather private initiative, somewhat being 'pulled' to the market by the lure of commercial goods in the form of processed foods and consumer items. With respect to the timing of market access, we find parallels between both communities that have to do with the construction of a road some three decades ago. This spurred the arrival of commercial goods to the region, generating also a higher influx of these items into the communities, either via itinerant traders or direct purchase from the market. Another similarity is related to the accumulation of debts that demands an individual or a whole family to sporadically increase market participation. In Nalang, a system of loan-giving developed through private initiative has drawn many producers into debt, since crops are repeatedly sold before they are harvested. Similarly, producers in Campo Bello are all too often indebted with exploitative river traders. At times, debts accumulated with traders before the harvest period oblige the people to sell a large amount of their produce. As a consequence, a more large-scale consumption of market goods is seemingly prevented in the absence of high income accumulation (see Grünbühel et al. 2007). On an interesting footnote, those commercial goods obtained from the market have rather intensified, or, at the least facilitated, general subsistence production in both communities.

6.5.3 Direct State Interventions and Other Development

Finally, to what extent can direct system interventions trigger a regime to change? SangSaeng's direct intervention experience is embedded in a larger regional state-induced effort that has aimed to integrate northeast Thailand into the nation-state. Until the 1960s, northeast Thailand was a sparsely populated back-country area of relatively little general importance. Inside the community, the state has directly intervened in the construction of a partly paved road and a gravel path, communication projects, educational and cultural assimilation programmes. The high material turnover of 18.6 t per person in part reflects these widespread infrastructural developments. Despite some political resistance, national integration efforts have generated also socio-cultural changes at the local level. Today, the people of Sang Saeng respect the King of Thailand as the patron of their country, they perceive their local language as a dialect not to be used in official settings, and accept their low social status in Thai society. State intervention in the political sphere is confined to the village headman who mediates local concerns with the outside world.

State intervention programmes in Nalang are limited and only have a marginal impact on village politics and resource use patterns. The construction of the road, for instance, was enforced by semi-private logging companies rather than forming part of governmental development efforts. State induced land-use regulations were introduced recently, but they have not (yet) led to a transformation of local agricultural production. In Grünbühel et al.'s (2007: 166) view, 'the government appears to refrain from intervention as long as the community is able to produce sufficient food'. While this is the case for the time being, market demands for consumer goods are, however, on the increase. And with the need for more cash comes the need to intensify further the use of the available cultivable land and the (illegal) extraction of timber. Since such an undertaking inevitably leads to a drain of the local resource base, temporary or permanent out-migration will be the last resort available to the villagers.

Campo Bello has long been excluded from state intervention programmes. Apart from its relative remoteness, this has also got to do with the fact that the community is exclusively inhabited by indigenous Tsimane' families, who until some decades ago, did not enjoy any rights as Bolivian citizens. It was only with the foundation of the Great Tsimane' Council in the late 1980s, that this ethnic group gained some political voice. In the early 1990s, increasing exposure to national society urged the Tsimane' to elect community leaders. Following the national system, villagers chose their political representatives periodically as mediators between themselves and the outside world. However, the limited knowledge of Spanish and the generally low status of the Tsimane' ethnic group generally prevents more than just sporadic state intervention. Among the few examples was the construction of a concrete school building. Other direct interventions include a health post built in a communal effort as part of a larger NGO-financed health programme. Upon last visiting the community, the building was used as an additional school building to grapple with the growing numbers of pupils. Another external effort was the construction of a lodgement for biosphere park rangers on the outskirts of the community in 2006. These infrastructure developments have hardly triggered any land cover changes, but contributed to an enormous increase in material consumption. This is also reflected in the data. Even though we find the weight of materials in Campo Bello to be the lowest of all the four cases (1.4 t/cap), Table 4.4 presents an 11.3% increase in material consumption between 2004 and 2006. This increase in uva grass, cement and gravel is largely the result of these outside interventions rather than of private initiatives.

Beyond the local level, state-induced colonisation programmes to the region have fuelled encroachment by highland settlers and affected the Tsimane' in two ways: on the one hand, competition for land has increased, pushing the indigenous people into more marginal lands.[18] On the other hand, many newly established settlers also engage in rice cultivation, pulling down market prices. At the same time though, the influx of outsiders to the region has also increased short-term wage labour opportunities for the villagers as well as prompted local trade relations with nearby farms and cattle ranchers.

As for Trinket, state intervention has played an important role in triggering transition processes. The Nicobar archipelago became part of independent India in 1947 and in 1956, the *Protection of Aboriginal Tribes Regulation* was launched in an effort to control entry to the Nicobars and offer protection to the islanders. Since the 1980s, the so-called 'economic upliftment' programmes have gradually been introduced on the islands in an attempt to integrate the Nicobarese into mainstream Indian society. On Trinket, welfare schemes introduced by the local administration include a school, solar lighting and a powerhouse. These infrastructure inputs are reflected also in the material weight that accounted for 8.3. t/cap in the year 2000. Singh (2003a: 118) also calculated a 10.2% increase in material stocks in just 1 year. Government housing and development schemes are also responsible for additional land cover changes. In 2001, the land area under non-traditional construction augmented by a considerable 48% as compared to the year before. These infrastructure development efforts generated an influx of labourers from elsewhere, who would gradually establish small enterprises and install small markets for the trade of copra in exchange for consumer goods. This increased connection with the outside world has led the locals to seek a lifestyle based on higher market demands that can only be met by increasing copra production and accruing arrears with the traders (Grünbühel et al. 2007: 167).

6.6 Conclusion: Lessons from Contemporary Transitions

From the outset of this chapter, the intention was to compare four contemporary local transitions and, with the help of the intensification and sociometabolic regime theory, find any common pathways in terms of their biophysical performance and driving forces that trigger such transformations. As far as these four cases even permit a tentative conclusion, I consider an examination of land and labour efficiencies to best recapitulate that which has been discussed so far.

Out of the four cases, Trinket shows by far the highest efficiency rates in terms of land and labour and has been referred to as almost 'paradisiacal' (Grünbühel et al. 2007: 156), with extremely low human labour inputs, and food staple yields more than five times higher than those of SangSaeng, three times those of Nalang and twice those of Campo Bello. The people of Trinket invest on average five times less than the people of Nalang in the reproduction of their economic system, while

[18] These interventions largely took place before the Bolivian State officially recognised the Tsimane' Territory in 1990, granting legal land titles to various indigenous groups in the region.

horticultural activities require a minimal daily labour input that is indeed 25 times lower than that of Campo Bello. This also implies that besides low work-loads, the locals of Trinket also exert a considerably lower pressure on their land resources, which is also reflected in their low appropriation level of net primary production. For the time being, Trinket is still far from reaching land and human resource limits. Rich forest and marine resources coupled with a tremendously low population density all account for the rich and diversified diet enjoyed by the local population. It seems that the system has managed to merge the best of both worlds, combining the 'leisurely' lifestyle of traditional foraging societies (low labour input to meet subsistence needs in combination with a rich diet) with a well-organised form of using the market. While they still largely follow traditional production and exchange patterns, the sale of copra allows them to meet modern-day needs. In return for copra they obtain various products, among them high-quality rice, for the production of which they would have to accept a great deal more labour time and effort as invested for copra production. This current security net based on the state-induced subsidies that ensure an efficient exchange rate between copra and rice, however, makes the local economy extremely vulnerable and dependable on externalities. If the government changes its subsidy policy as a consequence of a global increase in diesel prices, the locals would have to intensify their labour inputs in horticultural activities, for which, however, they still have a large safety margin available. With an increase in these activities, colonization activities will equally intensify; again, with still room to do so. Copra production, on the other hand, would likely diminish as transport inputs would no longer be readily available.

The case of SangSaeng shows a different efficiency profile, since both, land and labour resources are fairly limited. Productive efficiency can hence be achieved through the use of fossil fuel inputs only, which has led to an increase in environmental costs (e.g., the use of chemical fertiliser has degraded local soils) and a material burden that is more than twice as high as that of Trinket. In terms of human labour time, however, we can observe that the use of fossil fuels has probably decreased the burden on the people. While we see that, for the time being, the system handles its limited natural resource base by out-migration, what does the future hold? One of the future options would be a further increase in fossil fuel-based industrial irrigation agriculture. To take this thought further, an increasing use of fossil fuel inputs would very likely replace, at least to some degree, human labour inputs, which would, in turn, trigger more migration to urban hubs in search of wages to pay for additional inputs. A transition to fossil fuel-based irrigation agriculture would also be facilitated through fairly easy access to external markets. Notwithstanding, the major constraint on this development is the limited availability of water in the community.[19]

Nalang and Campo Bello are both on the lower end of the efficiency scale, with Nalang's showing substantially higher inefficiency levels. It is the system with the

[19] In 1998, per capita water consumption in a household in Sang Saeng amounted to 107 l. Total annual water consumption amounted to 8,868,789 l, out of which around one third were used for agriculture (Grünbühel et al. 1999: 20).

most human labour inputs, which are in fact almost three times higher than those of Campo Bello, but the least labour productivity. And this, quite interestingly, is even the case despite the use of animal labour and some, albeit low-level, fossil fuelled technologies for food production. These inputs increase the local peoples' dependency on external markets. One possible justification for describing the low efficiency phenomenon is the fact that the villagers do not feel the need for more efficiency. There is still enough room before Nalang reaches a deadlock in terms of land and labour. Subsistence needs are met with relative ease and with a low population density, Nalang's inhabitants still face little difficulty in feeding themselves. And, there is yet little incentive to change their modes of production. Not only does the community not face any acute pressures, but it seems that in the light of inadequate transport facilities and hence market opportunities, there is in fact not so much temptation to be more efficient. The community's development path would in all probability have been different had Nalang had as easy access to markets as Sang Saeng. Quite likely, they would have already started cash crop rice production. For the time being, market production is limited to dry-season small-scale cucumber cultivation. The community still has the chance to increase its population size and optimise its resources without having to undergo drastic changes in the production system.

Bearing some resemblance to the situation in Nalang, the people of Campo Bello also seem to lack an 'incentive' to invest more time in agriculture and, in doing so, increase their productive output. For the time being, time spent hunting, gathering and fishing remains a vital element of the local subsistence system. Shifting cultivation is likely to remain the only agricultural land use strategy the people can opt for, at least as long as no external inputs are acquired on a large scale to overcome the ecological constraints of poor soil fertility. Extensive swidden cultivation allows the residents to invest comparatively little labour time to meet subsistence and market needs. And there is yet another reason why the people of Campo Bello have little motivation to change their production strategies: they are well fed and enjoy a varied diet. People do not depend solely on rice as their main staple. About half of the entire rice production is sold on the market in exchange for tasty pasta noodles, an imported staple that is becoming increasingly popular among the Tsimane'. At the moment, it seems that the community of Campo Bello follows diversified production strategies with relatively little impact on the local resource base, a typical small-holder strategy that ensures food security (see Netting 1993). On my last visit to the community in 2008, quite a few people expressed their desire for chainsaws to topple more trees with less time and energy effort. This would entail an opening of rice fields from primary forest, which is said to host particularly suitable soil for rice production. While the village still has almost half of its territory covered in primary forest, the problem with chainsaws would lie much more with the increasing dependency on fossil fuel inputs from the market to ensure the running of their new agricultural devices. The options to generate more money would then be threefold: an increase in wage labour, an increase in handicraft sales (less likely since mainly women are engaged in this kind of activities), or, and probably most likely, an intensification of rice and plantain production. The latter, however, would mean substantially more labour demands on women and children.

From this comparative approach on local transition dynamics, we have learned some interesting new insights. First of all, Trinket serves as an example that market exposure, even over long periods, does not automatically lead to a change in local production patterns and, as a consequence, an intensification of land use. Campo Bello and Nalang both demonstrate how access to markets may in fact enhance subsistence production and the productivity of time (through the use of synthetic fishing nets, for example). Another interesting insight is the fact that the impacts of state interventions, at least in Campo Bello and Nalang, have not substantially triggered the systems to move forward in their transition life-cycle, but rather had punctual impacts. Campo Bello though is of low state interest, not least due to its cultural composition as the indigenous Tsimane' are socially and politically discriminated. With Nalang, on the other hand, the low state interest seems to be more related to its geographical remoteness. Finally, the analysis of biophysical intensification patterns has also demonstrated that intensification theory, based on an analysis of preindustrial societies, still proves a useful frame for understanding the dynamics of contemporary agrarian societies. A comparison of Campo Bello, Nalang and SangSaeng indeed illustrated that the use of land and labour intensify when systems move from more extensive to more intensive agrarian production. Having said this, however, intensification theory alone would not be enough to help explain the biophysical dimension of contemporary transition dynamics. The first shortcoming is related to the driving forces that trigger a system to change. From the previous analysis we have learnt that there are in fact several dimensions that may interact simultaneously at different scales. While endogenous processes do have certain relevance, it is the exogenous processes which seem to have (at least in some of our cases) more fundamental impacts on transition dynamics. Fossil fuel inputs, even indirectly in the form of transport, for instance, may change system dynamics enormously. Industrialised technologies have the ability to increase labour productivity and hence overrule intensification theory, however, often to the detriment of the environment and increasing vulnerability to externalities (e.g., global fuel crisis). Therefore, the theory of sociometabolic transitions provides for new insights, as it points to a whole new array of exogenous forces that may instigate a system to move forward in its life-cycle (see also Fischer-Kowalski et al. forthcoming).

So to return to the question we asked at the onset of the chapter, whether we can detect a transition trend expressed in biophysical patterns of intensification, the answer would be a tentative yes. There is a certain trend for Campo Bello, Nalang and SangSaeng, who may somewhat be ordered sequentially along a continuum from a more extensive to more intensive modes of agrarian production. SangSaeng, however, also substantially includes industrial elements in its production system. Trinket's biophysical performance appears more of a hybrid, since it contains many features that are typical of hunters and gatherers, while at the same time comprising significant amounts of fossil fuel inputs and material stocks. Nonetheless, any conclusion efforts at this level can be at most preliminary. First, it would need many more cases, including sites from more temperate climates, to come up with any serious conclusions. But more importantly, the research question would need rephrasing, since what can be achieved so far is a static description of different local metabolic profiles rather than capturing transition processes. If finding common patterns remains indeed the core interest within the transition study paradigm, then efforts need to go into developing sets of biophysical indicators that indeed capture process dynamics rather than comparing single metabolic profiles only.

References

Bolhuis, E. E., & Van der Ploeg, J. D. (1985). *Boerenarbeid en stijlen van landbouwbeoefening*. Leiden: Leiden Development Studies.

Boserup, E. (1965). *The conditions of agricultural growth. The economics of agrarian change under population pressure*. Chicago: Aldine/Earthscan.

Boserup, E. (1981). *Population and technology*. Oxford: Basil Blackwell.

Carlstein, T. (1982). *Time resources, society and ecology, Volume 1: Preindustrial societies*. London: George Allen & Unwin.

Descola, P. (1996). *In the society of nature: A native ecology in Amazonia*. Cambridge: Cambridge University Press.

FAO. (2001). *World agriculture: Towards 2015/2030. A FAO perspective. Food and agriculture organization* (FAO). London: Earthscan.

Fischer-Kowalski, M., & Haberl, H. (2007). Conceptualizing, observing and comparing socioecological transitions. In M. Fischer-Kowalski & H. Haberl (Eds.), *Socioecological transitions and global change: Trajectories of social metabolism and land use* (pp. 1–30). Cheltenham, UK, Northampton, USA: Edward Elgar.

Fischer-Kowalski, M., Singh, S. J., Ringhofer, L., Grünbühel, C. M., Lauk, C., & Remesch, A. (forthcoming). Sociometabolic regimes in indigenous communities, in scale interactions with an industrial world context, and the crucial role of working time. A comparison of case studies. Submitted to *Human Ecology*.

Fukui, H. (1993). *Food and population in a northeast Thai village*. Honolulu: University of Hawaii Press.

Godoy, R., Reyes-Garcia, V., Huanca, T., Leonard, W. R., Vadez, V., Valdés-Garcia, C., et al. (2005). Why do subsistence-level people join the market economy? Testing hypotheses of push and pull determinants in Bolivian Amazonia. *Journal of Anthropological Research, 61*, 157–178.

Grünbühel, C. M., Schandl, H., & Winiwarter, V. (1999). *Agrarische Produktion als Interaktion von Natur und Gesellschaft: Fallstudie SangSaeng*. Vienna: Social Ecology Working Paper 55.

Grünbühel, C. M., Haberl, H., Schandl, H., & Winiwarter, V. (2003). Socio-economic metabolism and colonization of natural processes in SangSaeng village: Material and energy flows, land use, and cultural change in northeast Thailand. *In Human Ecology, 31*(1), 53–87.

Grünbühel, C. M., Singh, S. J., & Fischer-Kowalski, M. (2007). The local base of transitions in developing countries. In M. Fischer-Kowalski & H. Haberl (Eds.), *Socioecological transitions and global change: Trajectories of social metabolism and land use* (pp. 139–178). Cheltenham, UK, Northampton, USA: Edward Elgar.

Harris, M. (1977). *Cannibals and kings: The origins of cultures*. New York: Random House.

Hobbes, M. (2005). Material flow accounting of rural communities: Principles and outcomes in South East Asia. *International Journal of Global Environmental Issues, 5*(3/4), 194–224.

Krausmann, F., & Haberl, H. (2002). The process of industrialization from an energetic metabolism point of view. Socio-economic energy flows in Austria 1830–1995. *Ecological Economics, 41*, 177–201.

Lee, R., & De Vore, I. (1968). Problems in the study of hunters and gatherers. In R. Lee & I. De Vore (Eds.), *Man the hunter* (pp. 3–12). Chicago: Aldine.

Mayrhofer-Grünbühel, C. M. (2004). *Resource use systems of rural smallholders. An analysis of two Lao communities*. PhD thesis, University of Vienna.

Netting, R. M. (1993). *Smallholders, householders: Farm families and the ecology of intensive, sustainable agriculture*. Stanford: Stanford University Press.

Polanyi, K. (1957). *The great transformation: The political and economic origins of our time*. Boston: Beacon Press.

Reyes-García, V. (2001). *Indigenous people, ethno-botanical knowledge and market economy: A case-study of the Tsimane' Amerindians in lowland Bolivia*. PhD thesis, University of Florida.

Sahlins, M. (1972). *Stone age economics*. London: Tavistock.

Sanchez, P. (1976). *Properties and management of soils in the tropics*. New York: Wiley-Interscience.

Schandl, H., & Grünbühel, C. M. (Guest Eds.). (2005). Southeast Asia in Transition. *International Journal of Global Environmental Issues*, *5*(3/4) pp 115–118.

Sieferle, R. P. (1997). *Rückblick auf die Natur: Eine Geschichte des Menschen und seiner Umwelt*. Munich: Luchterhand.

Singh, S. J. (2003a). *In the sea of influence: A world system perspective of the Nicobar islands*. Lund: Lund University.

Singh, S. J. (2003b). Transitions on an island society: A biophysical reading of society–nature interactions. In K. Deb & L. Srivastavan (Eds.), *Transitions towards sustainable development in South Asia* (pp. 82–197). New Dehli: The Energy and Resources Institute.

Singh, S. J., & Grünbühel, C. M. (2003). Environmental relations and biophysical transition: The case of Trinket island. *Geografiska Annaler*, *85B*(4), 187–204 (distributed by Blackwell: UK & USA).

Singh, S. J., & Schandl, H. (2003). Socio-economic metabolism in the Nicobar islands: Empirical research in society–nature interactions. In B. Benzing & B. Herrmann (Eds.), *Exploitation and over-exploitation in societies past and present* (pp. 169–184). Münster: LIT.

Singh, S. J., Grünbühel, C. M., Schandl, H., & Schulz, N. B. (2001). Social metabolism and labour in a local context: Changing environmental relations on Trinket island. *Population and Environment, 23*(1), 71–104.

Weisz, H., Fischer-Kowalski, M., Grünbühel, C. M., Haberl, H. Schandl, H., & Winiwarter, V. (1999, September/October). *Sustainability problems and historical transitions – A description in terms of changes in metabolism and colonization strategies*. Proceedings of the international conference, nature, society and history: Long-term dynamics of social metabolism, Vienna, Austria.

Wilkinson, R. G. (1973). *Poverty and progress: An ecological model of economic development*. London: Methuen.

Chapter 7
Local Visions for a Global Future

Abstract This final chapter sums up the main findings and emphasises the importance of starting to use the MEFA framework, or at least components of it, beyond the abstract of the classroom in development work. The main findings bring to the fore the interrelatedness of the biophysical features discussed in the previous chapters. Moreover, this chapter also briefly recapitulates the prevalence of common transition patterns as systems move through the transition life-cycle. In terms of drivers for change, this chapter also concludes that different forces are at play from higher scale levels, which all may have differing impacts on the dynamics of the local system.

7.1 Introduction

Through the lens of the local, we have accompanied the Tsimane' of Campo Bello in their daily rounds; mostly solitarily, we have seen them in a comparative light only toward the end of this journey in Chapter 6. Guided by the overarching aim to help modelling common development trajectories toward a sustainable future, this book has tried to contribute with the empirical case of a local indigenous society in transition. More specifically, its aims were twofold, yet intricately connected: to gain an in-depth understanding of the individual biophysical processes that occur at the local level, on the one hand, and to improve our understanding of transition dynamics, on the other hand. The first goal was based on the assumption that each system proposes its own solutions to universal problems, offering its own store of lessons to be learnt. The second goal was to add to the more ambitious aim of analysing the biophysical performance of transition dynamics. Embedded in the overarching research interest to discern common features of sociometabolic transition pathways, new findings are hoped to facilitate the guiding process of local systems under transition into a sustainable future.

Whether this book has contributed to these goals remains for the reader to judge. What is beyond doubt, however, is the innovative way this book has presented of looking at biophysical dynamics at the society–nature interface. Analysing material and energy exchange relations between a society and its environment call for rigorous

L. Ringhofer, *Fishing, Foraging and Farming in the Bolivian Amazon: On a Local Society in Transition*, DOI 10.1007/978-90-481-3487-8_7,

quantification methods whose sets of indicators may be regarded as signposts toward possible and impossible development trajectories. At the same time, I have also tried to give considerable weight to qualitative descriptions as a basis upon which to interpret individual findings in an accurate light. Methodologically, it is hoped that lengthy descriptions of data collection processes in one local system have provided some guidance for future students interested in such empirical undertakings. But my vision goes beyond classroom teaching, as it is for science to also guide development work in the field. At some instances throughout this monograph, we have come across the short-sightedness with which certain development projects are (still) implemented. Be it the remnants of a telephone setting or quasi-abandoned steel latrines in Campo Bello, some interventions simply do not hold what they may well-intentionally have set out to deliver. At the same time though, we have also learnt of development interventions that have had, at times probably unintended, impacts. The set-up of a mattress factory in San Borja, for example, has generated a fairly intensive exploitation of balsa seed pods, the raw material required for the stuffing of mattresses. While certainly a positive development in terms of income generation especially for Tsimane' women, one of the inadvertent consequences has been a growing trend to fell balsa trees in order to reach all seed pods. Balsa wood, however, is a highly appreciated resource among the Tsimane' and remains one of the most significant raw materials for housing and artefact construction. From these concrete examples we may possibly progress to a more general assertion: the promotion of certain development interventions (e.g., cash crops) possibly will provide a lucrative goal, at least in the short run. In the long run, however, they may not promise sustainability. Some interventions have indeed undesirable consequences ranging from increased labour demands on certain sex and age groups, a growing dependency on external inputs (e.g., fossil fuels), or, as we have seen, detrimental effects on the environment. With these dynamics in mind, I believe that the MEFA framework, or at least parts of the toolkit presented here, may offer development practitioners a viable monitoring instrument for measuring biophysical changes during the project cycle of a development intervention.

7.2 What Have We Learnt?

For now though, it remains to recapitulate the main findings that have gradually been established along the way through the previous chapters and attempt to draw a few lessons. Let us begin with the first aim guiding this monograph, namely to enhance our understanding of the biophysical performance of the system of Campo Bello.

If we take a transition to be a development that tends to evolve gradually within a given state and, at critical thresholds, enters a period of transformation ('take-off' and 'acceleration' phase), which may be chaotic and turbulent until it stabilises again in a revised state with qualitatively different features (see Raskin et al. 2007), then we find Campo Bello as a kind of hybrid between two parallel transitions. Some, albeit few, of its metabolic features are fairly close to what has been termed the hunter and gatherer sociometabolic regime. The system lives on extremely low material and energy inputs and average nutritional intakes are somewhat comparable to those calculated for past hunter and gatherer regimes. For its economic and social features, the system's metabolic profile undeniably falls within the agrarian sociometabolic regime. Population density is similar to that calculated for past agrarian regimes and the growing importance of agriculture is somewhat reflected in the high export rates to other social systems. Yet contrary to past regimes, the difference here is the insertion of industrial elements (i.e., kerosene and diesel) in peoples' daily livelihoods. Although this inclusion is still fairly marginal in terms of quantity, the system has entered a 'take-off' phase toward industrial production modes. While this implies a still early stage within its transition process to the industrial regime, currently changing tastes among the local people, coupled with the existence of cash and growing market demands, are all likely drivers for the system to enter an 'acceleration phase' of its transition toward an industrial regime in the not so distant future.

But the biophysical set-up of Campo Bello is not so unique after all, if we take into account that around 60% of today's rural smallholders, especially in the countries of the south, still sustain themselves from agricultural practices, foraging, fishing and small-scale trade. In terms of land, we talk about two-thirds of the world's total land area (see Fischer-Kowalski and Haberl 2007). These systems, though with different speed levels and intensities, are all in midst of one, or sometimes even two parallel transitions, often experiencing similar dynamics to those found in Campo Bello. Yet contrary to past sociometabolic transitions, today's transitions happen at a much faster pace. With the injection of cash, people can readily obtain fossil fuel based technologies that may cause enormous land cover changes and other ecological impacts. This industrial mode may promise more in terms of higher living standards, but these strategies are certainly not sustainable. As global material and energy resources are slowly diminishing, they would simply not suffice to cover the arising material, energy and land consumption needs that would occur if these rural systems are to undergo a similar transition process to that of past transitions from agrarian to industrial modes. In such a light, following the same pathway as the industrial north may simply not be an option for many, or probably most, of today's rural smallholder societies.

In such a light, which lessons can be drawn from the biophysical study of Campo Bello that may help the guiding process of local rural systems to move through a new kind of transition, a sustainability transition?

As it intricately connects to all other biophysical variables, population dynamics may serve as a good starting point. The number of people living in a system is in fact decisive for the kind and extent of biophysical material and energy exchange relations the society maintains with its environment. For pre-industrial societies, the amount of human labour available in a system is equally instrumental for the peoples' interaction with their environment. Some aspects of Boserup's (1965) theory serve particularly useful in exploring how population growth triggers the intensification of land use, which, in turn, at least in agrarian regimes, increases the demand for additional labour investments. The options for these societies to provide more labour are various and are strongly linked to socio-cultural practices (e.g., the acceptance of child labour) and the sexual division of labour. With regard to the system of Campo Bello, for instance, sexual labour allocation among children is not at all stringently imposed; a fact that can substantially relieve the indirectly productive work-load of the parents. Yet, the detailed time study in Chapter 5 has also shown that women bear the main brunt of increased labour. A woman's time inputs in the economic system accounts for about 47%, while the reproduction of the household system requires substantially more of a woman's time resources as compared to a man's. It may even be argued that this is because of the not so strict labour division in agriculture, since, in the absence of a man, women can take over in the productive sphere with only little help from others. Within the household, on the other hand, it would still be (largely) unthinkable for a man to engage in child care or cooking, both of which remain in the pure realm of a woman. So we see that high population levels have indeed an adverse effect on a woman's quality of life. With a hitherto annual growth rate at 3.77% and an upward tendency to be likely expected, this labour burden on women is likely to increase further, forcing them to reduce their time investments in other activities, most likely leisure, rest or sleep. In the long run, these developments may lead into a deadlock of social conflict and/or deteriorate the women's state of health (see also Fischer-Kowalski 2007). Further population growth will also lead to the degradation of land resources and a substantial increase in material and energy consumption levels.

Let me turn to the energy consumption of Campo Bello. As we observe an extremely low energy profile, Campo Bello's room for potential growth in energy consumption remains very high. The fact that all primary energy inputs for cooking derive from locally extracted biomass, the depletion of the local resource base for firewood will be the inevitable next stage (even if still at present, no scarcity of timber resources was observed). One exit strategy would be a shift to other cooking devices like gas ovens, which would not solely relieve the work-load on women and children who could then engage in other tasks, but as well substantially ease the burden on the local timber base. For the time being, however, this is unlikely to occur for various reasons, but mostly because people do not feel the immediate need to change their cooking practices. The community is still far from running into a shortage and, as long as this is the case, the system will still continue to use its traditional cooking sources before hitting the take-off phase to imported fossil fuels for food preparation. However, what historic transitions have shown us, once entering the take-off phase of a transition, a change toward the large-scale use of fossil fuels for food preparation will likely happen very fast (acceleration phase). The lack of felt needs is probably also the reason for Campo Bello's rather low conversion efficiency rates from primary to final and eventual useful energy. In the absence of large livestock, these low rates are mainly a result of the low conversion efficiency with which humans transform food into labour (see Smil 2006). For Campo Bello, the energy loss accounts for 68% from primary to final energy and 12% from primary to useful energy; a number that is similar to the energy metabolism calculated for Austria in 1830 (see Krausmann and Haberl 2002). The provision of process heat as a result of open fireplaces presents the lion's share of final energy and useful energy, accounting for almost 70% and more than 90%, respectively.

Another important insight is the inherent connection between the energy system and the use of materials, as the former shapes the latter. A low energy system based primarily on biomass (lack of conversion technologies, reliance on endosomatic energy sources only) can only sustain low material quantities, in terms of volume per capita and area. At the same time though, if these limitations are removed due to the insertion of fossil fuel based technologies (also indirectly through the provision of transport infrastructure), the quantity of material imports is bound to increase substantially as inputs are needed to sustain these technologies. The study of Campo Bello also brings to the fore how rapid the influx of imported materials may in fact occur. Imported construction material experienced a 12% increase in only 2 years. Likewise, the purchase of market commodities witnessed a 13% increase over the same period. With the rise of industrialised items, villagers are growingly becoming dependent on the outside world as the forest no longer (freely) provides for these necessary resources. We may argue that especially the influx of commodities is a direct consequence of outside intervention, as specific emergency relief aid was given to the local families in the immediate aftermath of a river flooding in 2006. While I believe that this emergency aid has not necessarily triggered a completely new way of material use in the local system, it has, however, undoubtedly accelerated the cultural acceptance and hence insertion of these industrial items in the peoples' daily livelihoods.

This point becomes very clear if we look at local food preferences. Although lacking the necessary longitudinal data, I consider the outside food aid given for months after the river flooding to have accelerated a change in local food preferences. During that period, breakfast plantains and manioc were largely substituted with fried bread served with heaps of sugar. While the intensity of flour, oil and sugar consumption became gradually less with these outside food subsidies coming to an end, it is still somewhat remarkable to observe how readily these items were inserted in the local dietary system. The lesson to be drawn from this observation is that socio-cultural changes may indeed happen quite rapidly. The growing popularity of pasta noodles, on the other hand, tends to be more a consequence of the generally higher status inferred to imported food items. Since its insertion in the local diet has been more incremental, past noodles are very likely to remain an important staple in the community.

As far as the second aim is concerned, which lessons can be drawn from the comparative approach of four local cases in Chapter 6? If we return to our first interest, namely to discern similar patterns of transition pathways, then the answer would be a tentative yes. While obviously many more cases would be needed to come up with any definite conclusions, we do see certain biophysical features that tend to augment and/or intensify as systems move through the transition life-cycle. This is in fact easier to detect in agrarian regimes, since, in line with Boserup (1965) and her followers, intensification patterns appear more incremental. In terms of colonizing efficiency, land use, human labour, and material and energy consumption, the comparative data largely corresponds to these theoretical assumptions. With the insertion of fossil fuel based technologies, however, a whole new dynamics comes into play. We have also looked at the driving forces that trigger a sociometabolic regime to change. Here we find that access to markets and direct development interventions from higher scale levels have differing impacts on local systems. To start with market access, I have found that communities are not unanimously pulled into the market economy but also use the market to enhance their subsistence production. Also, the four communities have opted for different production strategies to cope with the access to markets, either by intensifying the production of staple crops, the insertion of a new cash crop, or a change to a different staple crop variety. A related finding is the fact that a system's incorporation into the market does not necessarily require an intensification of land use, as the case of Trinket clearly illustrates. But Trinket also shows the direct impact of state intervention on the use of materials and energy sources. Likewise, state subsidies for copra production make the system highly vulnerable to national and international market developments. SangSaeng's metabolic profile and time use patterns also reflect the strong influences of regional development efforts. Nalang and Campo Bello, on the other hand, demonstrate a somewhat different dynamic, as state intervention has been fairly limited up until now. How would the impact be if transport infrastructure was built to connect the communities to regional market centres? Or, to what extent would the provision of medical services and/or further educational services influence population dynamics?

These concerns at the closing stages of this monograph show the inherent limits of the MEFA framework approach as it stands now. What it can offer so far is a socio-metabolic model that describes the flow of material and energy, land and human time

Fig. 7.1 Crossing the river, Campo Bello, 2004

use prevalent in a system at a certain point in time. What it still lacks though is the provision of a full systemic account of all the variables and implied processes (see Fischer-Kowalski and Haberl 2007: 17–18), and, closely related to that, indicators that capture process dynamics, which would naturally build on the description of these metabolic features we have discussed throughout the book. A final challenge that remains one of the most crucial to take up and develop further, is how to relate the large array of quantitative data to inform development programme and policy decisions. It is true that the study presented in this book enables an illustration of the biophysical possibilities and bottlenecks only towards viable future development pathways; yet in the end, how a desirable local future may be defined locally, remains to be decided among those immediately concerned. However, as I see the inherent role of science to guide this process, one necessary starting point would be for science to step down from the world of the abstract and feed these insights back to those actors, whose future may be directly at stake (Fig. 7.1).

References

Boserup, E. (1965). *The conditions of agricultural growth. The economics of agrarian change under population pressure*. Chicago: Aldine/Earthscan.

Fischer-Kowalski, M. (2007). *Ageing, time use and the environment.* Presentation at the workshop: Research foresight for environment and sustainability – megatrends and surprises, EEA Copenhagen.

Fischer-Kowalski, M., Haberl, H. (2007). Conceptualizing, observing and comparing socioecological transitions. In M. Fischer-Kowalski, H. Haberl (Eds.), *Socioecological transitions and global change: Trajectories of social metabolism and land use* (pp. 13–62), Cheltenham, UK, Northampton, USA: Edward Elgar.

Krausmann, F., & Haberl, H. (2002). The process of industrialization from an energetic metabolism point of view. Socio-economic energy flows in Austria 1830–1995. *Ecological Economics, 41*, 177–201.

Raskin. P., Banuri, T., Gallopín, G., Gutman, P., Hammond, A., Kates, R. W., et al. (2007). Great transition: Where are we. In Great Transition: The promise and lure of the times ahead. Retrieved May 12, 2009, from http://www.eoearth.org/article/Great_Transition:Where_Are_We

Smil, V. (2006). *Energy: Beginner's guide*. Oxford: Oneworld Publications.

Index

FSC
www.fsc.org
MIX
Papier | Fördert gute Waldnutzung
FSC® C083411

Kolibri 360 GmbH
Ferdinand-Jühlke-Straße 7
99095 Erfurt, Deutschland
produktsicherheit@kolibri360.de